The Essentials of
SHOPPER
TECHNOLOGY

The Essentials of SHOPPER TECHNOLOGY

SHOPPER TECHNOLOGY
INSTITUTE

Edited by John Karolefski - Executive Director

outskirtspress
DENVER, COLORADO

The opinions expressed in this manuscript are solely the opinions of the author and do not represent the opinions or thoughts of the publisher. The author has represented and warranted full ownership and/or legal right to publish all the materials in this book.

The Essentials of Shopper Technology.
All Rights Reserved.
Copyright © 2012 Shopper Technology Institute
v3.0

Cover Photo © 2012 JupiterImages Corporation. All rights reserved - used with permission.

This book may not be reproduced, transmitted, or stored in whole or in part by any means, including graphic, electronic, or mechanical without the express written consent of the publisher except in the case of brief quotations embodied in critical articles and reviews.

Outskirts Press, Inc.
http://www.outskirtspress.com

ISBN PB: 978-1-4327-8679-3
ISBN HB: 978-1-4327-8682-3

Outskirts Press and the "OP" logo are trademarks belonging to Outskirts Press, Inc.

PRINTED IN THE UNITED STATES OF AMERICA

Contents

Introduction vii

SECTION ONE: SHOPPERS 1

CHAPTER 1
Capitalizing on the Smarter Consumer 3

CHAPTER 2
Own the Future of Shopping 27

SECTION TWO: LOYALTY 39

CHAPTER 3
The Retail Arms Race 41

CHAPTER 4
Change Management by the Numbers: Devising a Strategy That Produces Results and Advocates 53

CHAPTER 5
Actionable Shopper Segmentation 65

SECTION THREE: ENGAGEMENT 75

CHAPTER 6
Shopper Metrics for Optimizing Retail Performance 77

CHAPTER 7
The Evolution and Application of Virtual Shopping: Past, Present and Future 93

SECTION FOUR: ANALYTICS 109

CHAPTER 8
The Evolution of Trade Promotion Management 111

CHAPTER 9
The Trend Behind the Spend.. 125

CHAPTER 10
Clearing Up Confusion about the Demand Signal
    Repository 133

CHAPTER 11
Using Virtual Environments to Measure Shopper
    Behavior 141

CHAPTER 12
Understanding 'Ready-for-Purchase'.................................... 157

SECTION FIVE: DIGITAL 167

CHAPTER 13
Achieving and Sustaining Planogram Compliance 169

CHAPTER 14
Is Your Brand Winning or Losing in the Coupon and
    Promotions Game?.. 175

CHAPTER 15
Better, Faster, Smarter Shopping Experiences.................... 181

CHAPTER 16
Making Sense of Social Media 189

Index 201

Introduction

Technology has dramatically changed the way consumers gather information, shop, amuse themselves, and connect with others. A decade ago, few of us would have imagined that websites, smartphones, tablets and social media would be so commonplace and influential in our everyday life. Where all of this will take us remains unclear, but there is no question that the pace of change is remarkable.

Retailers and makers of fast-moving consumer goods have their own revolution to deal with. They now compete in a world of Big Data. A competitive advantage is often determined by how much they know about their shoppers, and how best to use that knowledge to serve those customers better. That could be by offering customized promotions, a digital circular, or an enhanced loyalty card program. At the shelf, it should mean stocking the right product for the right shopper at the right time.

Trading partners can now engage and entertain shoppers in stores with digital screens and informational kiosks. They can measure shopper behavior with unobtrusive video cameras with the end game of enhancing the path to purchase. They can observe shopper behavior in sophisticated virtual environments.

What Is Shopper Technology?

There is a term for all of these solutions – Shopper Technology. It is broadly defined as tactics and applications that engage and motivate shoppers, analyze their behavior, and enable trading partners to improve their operations. Retailers are embracing these new solutions, which are sometimes deployed in partnership with manufacturers that have their own unique applications for brands.

The providers of these solutions have invested mightily in the future of this industry. Their sophisticated offerings are built on a foundation of thought leadership. This book is all about sharing that information and insight. There are chapters that outline the challenges facing merchants and marketers. Some chapters present the essentials of shopper technology, while others offer a deeper dive into these solutions and their implications.

Here is how the book is organized:

SECTION ONE: Shoppers

Chapter 1, *Capitalizing on the Smarter Consumer*, explains why companies need to understand and engage today's techno-savvy shopper. Serving these customers means listening and knowing them, as well as empowering them.

Chapter 2, *Own the Future of Shopping*, stresses that companies need to own the future of shopping to be successful in the increasingly competitive marketplace. Think of it as shopper marketing in a digital world.

SECTION TWO: Loyalty

Chapter 3, *The Retail Arms Race*, vividly sums up the challenge that independent and mid-tier retailers face in maintaining the loyalty of shoppers in a marketplace dominated by major chains leveraging powerful customer data.

Chapter 4, *Change Management by the Numbers*, explains why shopper centricity has been one of most significant trends driving retail growth. But the root cause of success or failure in a shopper-centric plan lies in effective implementation and change management.

Chapter 5, *Actionable Shopper Segmentation*, states that loyalty depends on the ability to identify groups of shoppers with similar needs and behaviors so that each group's distinct needs can be met. Shopper segmentation should be the foundation for how decisions affecting the shopping experience are made.

SECTION THREE: Engagement

Chapter 6, *Shopper Metrics for Optimizing Retail Performance*, outlines the reasons why the industry needs measurement tools that provide direct visibility into the shopper decision process and enable shopper-centered metrics for improving every retail touch-point. Marketers can understand opportunities, test innovation, and track results in real-time.

Chapter 7, *The Evolution and Application of Virtual Shopping*, presents virtual testing solutions to assist in identifying how consumers shop specific categories. It provides a framework for engaging consumers with products, planograms and communication strategies that best meet their needs and maximize their path to purchase.

SECTION FOUR: Analytics

Chapter 8, *The Evolution of Trade Promotion Management*, stresses that trade promotion remains a critical tactic for getting more shelf space, creating brand awareness, and helping to increase store traffic. Trade Promotion Management covers the entire cycle of trade promotions and related spending between CPG manufacturers and retailers.

Chapter 9, *The Trend Behind the Spend,* reveals insights into today's trade promotion and shopper marketing spending environment in consumer packaged goods, including a look into the future and how key tactics are changing the retail landscape.

Chapter 10, *Clearing Up Confusion about the Demand Signal Repository*, outlines the process in simple terms. DSR integrates and cleanses demand data with internal data to provide business users with alerts and reports that will pinpoint areas of the business that require immediate attention.

Chapter 11, *Using Virtual Environments to Measure Shopper Behavior*, explains why virtual shopping is an effective tool for manufacturers and retailers. It delivers an accurate representation of in-store performance by observing shopper behavior, rather than relying on asking shoppers what they would do.

Chapter 12, *Understanding 'Ready for Purchase,'* introduces RFP as a methodology that delivers a new standard of retail performance. To reduce out of stocks, the strategy monitors POS transactions and applies pattern recognition in real time to determine what items are ready for purchase – and which are not – from the shoppers' point of view.

SECTION FIVE: Digital

Chapter 13, *Achieving and Sustaining Planogram Compliance*, discusses the four frustrating factors that are common denominators when retailers can't seem to get – and keep – their stores compliant. The solution to this long-standing and nettlesome issue calls for looking at data in a new way.

Chapter 14, *Is Your Brand Winning or Losing in the Coupon and Promotions Game?*, examines the five factors behind the proliferation of digital promotions today. Three suggestions are presented for brands and retailers to take advantage of these promotions.

Chapter 14, *Better, Faster, Smarter Shopping Experiences*, presents a digital solution that simulates retail settings realistically inside immersive life-like 3-D environments to better imagine, validate and deploy optimum shopping experiences. The focus is on the store shelf.

Chapter 16, *Making Sense of Social Media*, presents the results of a national study of consumer packaged goods manufacturers. It reveals how companies are using social media in brand marketing strategy and for customer engagements. And it reports on early attempts to measure social media.

Closing Thoughts

Because of Big Data, digital innovations, and an evolving path to purchase, the job of merchants and marketers often seems daunting. So consider this book a way to cut through the clutter as technology roars on, competition heats up, and shoppers become more demanding. Coming to grips with technological change may be challenging, but it is the key to success.

I'd like to thank the sponsoring members of the Shopper Technology Institute (STI) that contributed informative and insightful chapters to this book. I'd also like to acknowledge and thank Linda Winick, STI's Director of Operations, whose tireless and meticulous work made the production of this book possible.

The mission of STI is to inform and educate retailers and manufacturers about the new and established solutions that impact shoppers directly or indirectly. Among STI's communication vehicles: a twice-monthly newsletter, webinars, podcasts, the annual LEAD Marketing Conference – and this book. It is our first, but won't be our last.

John Karolefski
Executive Director
Shopper Technology Institute
www.shoppertech.org

SECTION ONE
SHOPPERS

CHAPTER 1

Capitalizing on the Smarter Consumer

By Melissa Schaefer

Today's consumers are more connected, vocal and demanding than ever. In fact, technology is a given in the smarter consumer's life. But what other forces are driving shoppers today? Findings from our survey of more than 30,000 consumers in 13 countries tell us that consumers want to be heard, known and empowered. Savvy retailers will listen to today's smarter, connected consumer and leverage what they learn to empower their customers to shop when and how they want.

In 1911, women wore ankle-length skirts, men wore three-piece suits topped with bowler hats, and horses were the most reliable "mobile devices." The department store was in its heyday, service was paramount, and the customer was "always right." The most advanced retailers used tabulating machines made by the newly formed Computing-Tabulating-Recording Company (C-T-R), but consumers still had to pay for their purchases with cash.

Fast forward 100 years, and the picture is very different. Women wear trousers as often as they wear skirts; the waistcoat and bowler hat have been consigned to history; web-enabled mobile devices are ubiquitous; and consumers are as likely to shop online as they are in department stores. They can call on a vast digital network of friends and fellow consumers to help them decide what to buy. And they can

4 | ESSENTIALS OF SHOPPER TECHNOLOGY

pay for their purchases using chip-and-PIN systems supported by IBM (which C-T-R became) technologies.

Thanks to technology, consumers have been getting smarter and smarter. But have retailers kept up with them? The results of our latest consumer survey suggest not.

In 2009, IBM surveyed over 30,000 consumers in three mature and three growth economies to find out how technology is transforming the way people shop. Through this study, "Meeting the Demands of the Smarter Consumer," we discovered that consumers are more connected, more vocal and more demanding than ever before.[1]

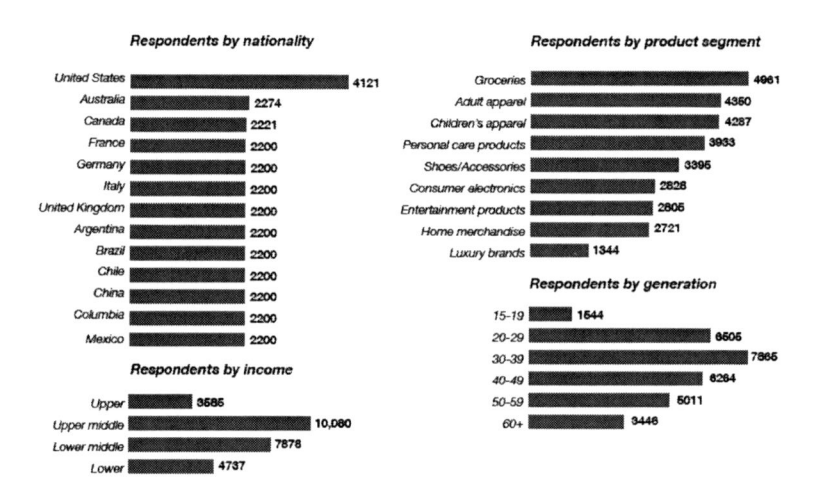

Respondents by nationality

Country	Count
United States	4121
Australia	2274
Canada	2221
France	2200
Germany	2200
Italy	2200
United Kingdom	2200
Argentina	2200
Brazil	2200
Chile	2200
China	2200
Columbia	2200
Mexico	2200

Respondents by income

Upper	3585
Upper middle	10,080
Lower middle	7878
Lower	4737

Respondents by product segment

Groceries	4961
Adult apparel	4350
Children's apparel	4287
Personal care products	3933
Shoes/Accessories	3395
Consumer electronics	2828
Entertainment products	2805
Home merchandise	2721
Luxury brands	1344

Respondents by generation

15-19	1544
20-29	6505
30-39	7865
40-49	6264
50-59	5011
60+	3446

Home merchandise refers to stores in which consumers shop for TVs, games, appliances, hardware, home décor or home furnishings.
IBM Institute for Business Value.

But we realized that technology was only one of the forces determining how shoppers behave and that other factors, like household demographics and socioeconomic shifts, might be equally important. So we decided to delve more deeply into the mind of the consumer. We wanted to know what motivates customers, who influences customers, how the shopping process is changing and, ultimately, how retailers should respond.

In late 2010, we consulted another 30,624 consumers to find out what they really think when they go shopping. We also surveyed more than 5,000 consumers on growth trends, including changes in the family unit and attitudes toward money.

The story that unfolded is complex. It shows that smarter consumers want retailers to listen to them, know them and empower them – at the same time that these consumers are becoming more difficult to know. What you see isn't necessarily what you get, and traditional strategies for finding out about customers certainly won't reveal the whole truth. It's essential to look more widely at information both inside and outside the retailer's enterprise.

So what, exactly, did we learn?

The consumer is digital. Smarter consumers take technology completely for granted. Nearly half the people we surveyed are eager to use two or more technologies to shop. Younger consumers are particularly keen – and teenage "twitterati" have now escaped the parental leash.

The household is "virtual." Many consumers are shopping for a much wider range of family members, as the number of mutigenerational households rises. Thanks to the Internet, consumers can easily shop for adult parents who may or may not live nearby.

Incomes and shopping attitudes are diverging. Between a fifth and a quarter of all consumers search for sale goods and only buy what they need, no matter how affluent they are or how optimistic they feel about their financial future.

Smarter consumers listen to their families and friends first. The vast majority of consumers talk to relatives and friends or read independent reviews when they want to know more about a product. Only 18 percent rely on retailers and manufacturers.

Smarter consumers shop differently. They use technology to start and stop the shopping process and take days or even weeks to complete the various steps in the process, instead of shopping in a continuous linear flow.

Smarter consumers want to be served. They want to shop as easily and conveniently as possible because they already know which products and brands they need before they enter the store. They don't want retailers to advise them because they have formulated their opinions by consulting family, friends and strangers.

Smarter consumers want to be known. They want a personalized shopping experience tailored to meet their needs and preferences. In fact, it's their top priority – as it was last year as well.

Smarter consumers want to feel empowered. They want to use mobile technologies to make the shopping process easier and more pleasurable. And they want to take possession of the purchases they make in the ways they choose.

Social media channels are a rich source of information and influence. Nearly half of all consumers who follow a brand on social media submit their own comments, and nearly two-fifths are more loyal to brands they have engaged with online.

Listening to consumers and acting on what they say is profitable. More than half of all consumers who follow a brand on social media spend more with retailers with which they have interacted positively online. And nearly 13 percent say the increase is significant.

In short, smarter consumers are harder to "read" than their predecessors. They are also more cautious about spending their money and more resistant to marketing because they have other ways to get the information they require. So retailers have to become smarter, too. Smarter retailers have to listen carefully, leverage what they learn, and help consumers shop when they want – as they want.

A Digital Shopping Universe

The number of Internet users has doubled from 2005 to 2010 – from approximately 1 billion to 2.1 billion.[2] The number of mobile phone subscriptions soared from just over 2 billion in 2005 to 5.3 billion in 2010, with smartphones accounting for nearly 20 percent of all mobile communications devices sold in 2010.[3] And social networking has taken the world by storm. More than 500 million people stay in touch with their friends on Facebook, and more than 175 million "tweet."[4]

The digital universe is transforming the way in which consumers interact, both with retailers and with each other. It is also giving them an unprecedented amount of power. For example, when Proctor & Gamble's "Smell like a man, man" campaign went viral, attracting 140 million views on YouTube and almost 120,000 followers on Twitter, sales of Old Spice Body Wash jumped 55 percent.[5]

Smarter consumers carry enormous clout. But who, precisely, are they? Our latest shopper survey aims to answer that question and to shed light on what retailers can do to earn a bigger share of the smarter consumer's spend (see sidebar on next page, *Methodology*).

Methodology

In October 2010, the IBM Institute for Business Value conducted an online survey of 30,624 consumers living in seven mature economies (Australia, Canada, France, Germany, Italy, the United Kingdom and the United States) and six growth economies (Argentina, Brazil, Chile, China, Colombia and Mexico). They represent every age and income group. (The purchasing power of the same sum of money varies dramatically among different countries. We therefore divided respondents into four income brackets, using the average income in each country as our midpoint.)

We classified respondents according to the product categories in which they frequently shopped (rather than by retail segment, as we have done in previous years) because we recognize that a single retail segment may cover more than one product category. Figure 1 shows the composition of the survey population by nationality, generation, income group and product segment.

We analyzed the results using various statistical techniques, including maximum difference (Max Diff) scaling – where respondents are asked to compare different statements and pick the statement that has the most and least influence in each set. This method mimics the way in which consumers shop for goods in real life.

We also surveyed 5,188 consumers in three mature economies (Australia, Germany and the United States) and two growth economies (China and Mexico) to determine how their attitudes toward shopping, incomes and purchasing have changed. We viewed the data through three different lenses: gender, age and income group.

The smarter consumer is different.

The Smarter Consumer Is Different

Consumers are connected. Technology is completely entrenched in the life of smarter consumers. They are comfortable using the Internet, mobile technologies, in-store kiosks and digital TV to browse for and buy goods. Last year, we identified a core group of people who were willing to use two or more such technologies to shop; we called them "instrumented" consumers. This year, 49 percent of respondents are instrumented – a 36 percent rise in 12 months. The number of shoppers who are currently not willing to use any technologies has also fallen to just 14 percent (*see graph*).

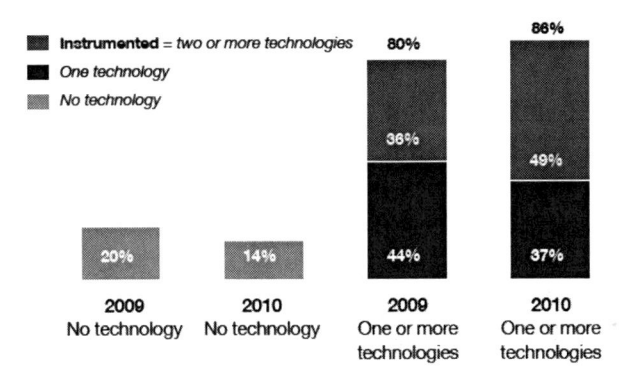

■ Instrumented = *two or more technologies*
■ One technology
▩ No technology

	2009 No technology	2010 No technology	2009 One or more technologies	2010 One or more technologies
Instrumented			36%	49%
One technology			44%	37%
Total with tech			80%	86%
No technology	20%	14%		

Source: IBM Institute for Business Value.

The Internet and in-store kiosks remain the most popular options: 75 percent of all consumers are willing to shop on a retailer's Web site, while 39 percent are willing to use in-store kiosks – a year-on-year increase of 10 percent. But interest in digital TV and mobile technologies is climbing even faster. The number of consumers who are ready to use digital TV has risen 41 percent (from 17 percent to 24 percent), and the number of consumers who are ready to use mobile technologies has soared by 92 percent (from 13 percent to 25 percent).

The surge of interest in shopping via digital TV – i.e., purchasing products by pressing a button on a remote control – is particularly noteworthy. It reflects the way in which the Internet, video and social media are converging, with the development of new devices that enable users to view content from iTunes, YouTube and the like on a high-definition widescreen TV. When Apple launched the redesigned Apple TV in the United States in September 2010, for example, it sold 250,000 units in 18 days.6 The digital TV shopping channel will become even more important for retailers in the future, as the line between home computers and TVs continues to blur.

Consumers in the growth markets still lead the charge in instrumentation, as they did last year. More than 60 percent of those living in Brazil, China, Chile and Mexico are willing to use two or more technologies to do their shopping, compared with just 30 percent of those living in France (*see graph, "Percentage of instrumented consumers by country"*). This is consistent with the fact that new technologies – such as mobile phones – often spread more rapidly in growth countries with relatively weak infrastructures because they can be used to overcome such limitations. But two other factors may be pertinent as well.

Percentage of instrumented consumers by country

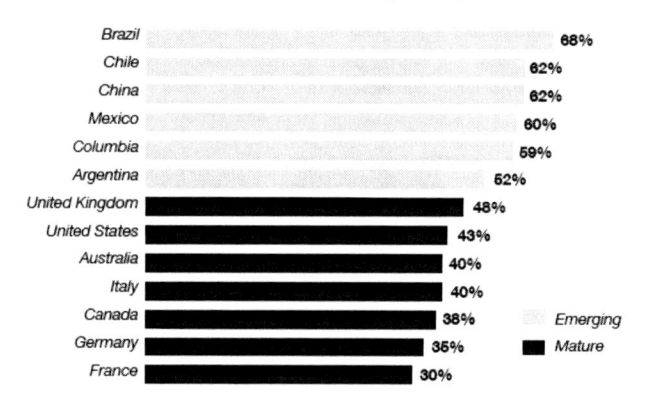

Country	%
Brazil	68%
Chile	62%
China	62%
Mexico	60%
Columbia	59%
Argentina	52%
United Kingdom	48%
United States	43%
Australia	40%
Italy	40%
Canada	38%
Germany	35%
France	30%

Emerging / Mature

Source: IBM Institute for Business Value.

The mega-cities of Latin America and Asia rely on public transportation to keep their citizens moving.7 In the mature world, by contrast, private transport is more common. Indeed, only 4.9 percent of U.S. citizens use mass transportation to get to work.8 Consumers who spend a lot of time on the bus or train may be turning to mobile technologies to help them use that time as productively as possible. Mobile phone tariffs are also relatively high in some European countries. The weighted average cost of a monthly mobile phone subscription is 60 percent higher in Germany and 107 percent higher in France than it is in the United States, for example.9

Younger consumers are likewise very keen on using technology to enhance the shopping experience. In fact, the "digital babies" have seized control. Last year, only 38 percent of 15- to 19-year-olds were instrumented – evidence, we concluded, that their parents still determined how they accessed and used technology. But mom and dad have apparently lost the battle, as 52 percent of teenagers are now using two or more technologies to shop (*see graph, "Percentage of instrumented consumers by age group"*).

Percentage of instrumented consumers by age group

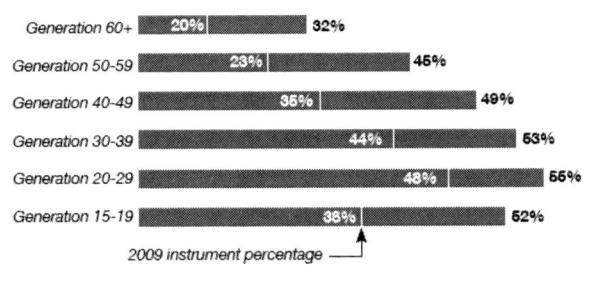

Generation	2009	Current
Generation 60+	20%	32%
Generation 50-59	23%	45%
Generation 40-49	35%	49%
Generation 30-39	44%	53%
Generation 20-29	48%	55%
Generation 15-19	38%	52%

2009 instrument percentage

Source: IBM Institute for Business Value

More surprisingly, perhaps, income does not seem to affect consumers' willingness to use multiple technologies in the shopping process. Shoppers in every income group are equally comfortable using two or more technologies, with the exception of those living in Australia, Canada, China and Colombia, where instrumentation is still more common among the more affluent.

However, technology is by no means the only factor reshaping the way in which consumers shop. Major demographic and socioeconomic shifts are also coming into play.

Consumers are shopping for extended families. A growing number of people are shopping for a much wider range of family members. This is partly because the family unit is changing; 25 percent of respondents have parents, adult children or grandchildren living with them (see sidebar, *The Return of the Multigenerational Household*). But it is also because many consumers are making purchases for relatives who don't even share a roof. More than 30 percent of respondents regularly buy clothing, groceries, consumer electronics, entertainment products and personal care products for their parents, whether or not they live together.

The person who makes the purchasing decisions is now making them for a "virtual" household that extends beyond the four walls of the traditional home. And this consumer wants new services specifically to cater to the needs of his or her aging parents. Sixteen percent of consumers want food retailers to deliver meals and groceries, for example, while 14 percent want consumer electronics retailers to provide installation and support services.

The Return of the Multigenerational Household

Multigenerational households have long been the norm in the growth world. In Latin America, for example, one in every four homes includes at least one older adult and more than two-thirds of older adults live with their grown-up children.[10] But the number of multigenerational households is now rising in the mature countries, too, as a result of various socioeconomic factors – including greater longevity, later marriage and higher unemployment rates.

An estimated 49 million Americans (16.1 percent of the total population) live in homes comprising three or more generations – up from 42 million in 2000.[11] The situation is similar in Western Europe, with 15 percent of all senior citizens with adult offspring living in the same household. Another 34 percent live within a kilometer of their children. Geographical proximity between different generations is particularly pronounced in Mediterranean countries like Italy and Spain

Prosperity is no guide to purchasing patterns. Incomes and shopping attitudes are simultaneously diverging. At one stage how much people earned was a good guide to how they shopped. That's no longer true. The majority of the consumers we surveyed are reasonably confident, or even optimistic, about their future income. Forty-six percent expect their income to stay the same for the next five years, while 24 percent expect it to increase by at least 20 percent – and these high hopes are shared by high and low earners alike.

But when we asked respondents about their shopping attitudes, the three most frequent responses were: "I only buy what I need" (20 percent); "I search for sale items" (19 percent); and "I wait longer to make a purchase" (16 percent). In other words, consumers are more careful with their cash, regardless of how affluent they are or how confident they feel about the future.

Further proof that consumers are pulling in their horns comes from the fact that, in the month preceding our survey, 41 percent of respondents changed their minds about buying something they had already put in their shopping baskets, and 37 percent abandoned an item in their online cart, primarily because they "didn't want to spend the money." Even when their motives differed, cost often played a part in the decision. Online shoppers in Brazil and China told us, for example, that they had terminated the transaction on discovering they could get the same product more cheaply elsewhere.

So smarter consumers are not about to return to pre-recession spending levels yet – a topic many retailers asked us about after last year's study. All the signs suggest the opposite: they won't be splurging in a hurry. The only countries where this sentiment differed are in the growth markets, particularly China and Brazil.

The biggest influences are those closest to home. Smarter consumers are not only more wary about dipping into their wallets, they are also more difficult to influence. Forty-five percent of respondents turn to friends and relatives, and 37 percent to external sources – either fellow consumers or independent experts – when they want to know more about products they are interested in purchasing. Only 18 percent trust retailers and manufacturers to give them an honest answer. This pattern is consistent across every product segment. Between 78 percent and 84 percent of consumers rely on their social networks when researching new products, irrespective of what those products are. That's not the only evidence of how consumers are "tuning out" retailers' messages. Fifty-nine percent of respondents won't even give retailers their primary e-mail address when asked. And if they do read a retailer's e-mails, they typically do so only to get loyalty program discounts or information about upcoming sales and promotions.

The Fragmented Shopping Process
Consumers have changed, and they have changed the shopping process, too. Instead of browsing through several stores, finding something and buying it in a continuous sequence, they use technology to weave in and out of the shopping process wherever and whenever they want. So what was once an uninterrupted flow is turning into a series of "moments": the moment of first becoming aware of a product, the moment of researching it, the moment of purchasing it and the moment of taking possession.

These moments may be separated by days or even weeks. Many consumers in Australia, Canada, the United Kingdom and the United States wait at least seven days between learning about a new item and purchasing it, for example. Consumers in the growth markets, by contrast, are more likely to purchase a product immediately or within a couple of days.

The shopping process has not only become more fragmented, it has also become more compressed. Time-starved smarter consumers can now go online to get the information they want in a matter of minutes, whereas once they might have spent an hour wandering through the store. Moreover, some of the biggest influences on smarter consumers during the moments when they first become aware of a product and research it are in their own hands. They are not simply responding to advertising and promotions; they are consulting their families and friends, using search engines and looking at mobile applications like ShopSavvy – media that are completely outside a retailer's control *(see diagram, "Greatest influences during product awareness and research).*

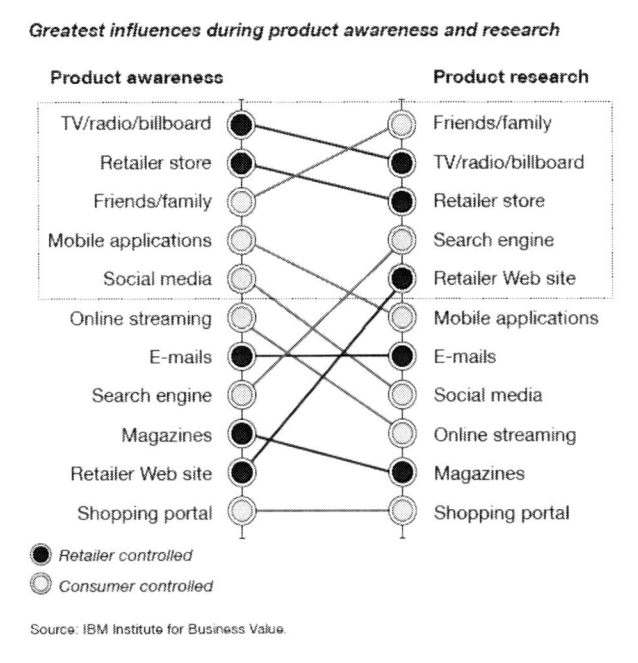

Greatest influences during product awareness and research

Product awareness	Product research
TV/radio/billboard ●	○ Friends/family
Retailer store ●	● TV/radio/billboard
Friends/family ○	● Retailer store
Mobile applications ○	○ Search engine
Social media ○	● Retailer Web site
Online streaming ○	○ Mobile applications
E-mails ●	● E-mails
Search engine ○	○ Social media
Magazines ●	○ Online streaming
Retailer Web site ●	● Magazines
Shopping portal ○	○ Shopping portal

● *Retailer controlled*
○ *Consumer controlled*

Source: IBM Institute for Business Value.

The retailer's window of opportunity for influencing the consumer has therefore become much smaller. It has minutes rather than hours to make a favorable impression, and it has to make that impression in a much "noisier" environment. A retailer is also less likely to successfully persuade the consumer to make impulse purchases because the intervals between the different steps in the shopping process provide plenty of time for reflection.

Serve Me, Don't Sell to Me
To sum up, it used to be relatively easy for retailers to identify their target customers, reach them and sell to them. These days, it's very hard. Smarter consumers are departing from their demographic and socioeconomic roots. They are also using technology to commandeer

the driving seat and control their own shopping experiences. When they enter a retailer's store or view its Web site, they usually know what they want because they have already talked to their families and friends and read the product reviews. They are the experts.

So how can retailers respond? Our survey shows that what smarter consumers really want is to be served, not sold to. They are telling retailers: listen to me, know me and empower me.

Listen to Me. Thanks to social networking in all its forms, including "tweeting," "tumbling" and "video hauling," consumers can converse with each other more easily than ever before. They can discuss their interests and experiences with different retailers, products and brands online – and many of them are doing precisely that.

Other research by IBM reveals that a majority of online users have social networking accounts, and almost half use media sharing accounts. The majority of them are casual participants; they occasionally post responses or post their own content. But only a small group – 5 percent – is responsible for most of the activity on social sites, nearly always responding to others' comments or authoring their own posts.[13] This consumer-generated content is a rich source of information for retailers.

Discerning the messages embedded in the digital arena isn't easy. Nor is it easy to participate in the dialogue. When consumers talk to a retailer, they expect something concrete in return. Most consumers don't follow a brand to "feel connected with it" or "join a community," as many of the retailers we have asked assume. On the contrary, they want discounts, trial offers or exclusive content – and they want them delivered in a way that is fun, fast and interactive.[14]

Nevertheless, listening carefully to customers provides invaluable insights into what they want – which products and services they desire, how they prefer to pay for those products and services, and how to deliver a better service. One top consumer electronics retailer has already learned this lesson – and it is partly why the company has survived when rival consumer-electronics retailers have disappeared (see sidebar, *Listening to the Buzz*).

Listening to the Buzz

When management at a top consumer electronics retailer started reading blogs and listening to social media, as well as encouraging local sales staff to report on local trends, it discovered that women did not like stores in a specific market for a very specific reason. Many of the women who shopped at these stores wanted to go straight to the store after dropping off their children at school. But the store didn't open until 10 a.m., so they either had to wait a couple of hours or come back later in the day. The company promptly changed the opening hours to 8 a.m., introduced several other changes to make the store more appealing to females, and publicized the improvements locally. The result? Sales soared.[16]

Know Me. Of course, listening to consumers is only the beginning. It's also crucial for retailers to show that they know their customers by providing them with a personalized shopping experience. Respondents told us that this is their top consideration when deciding where to shop and the area where retailers most need to improve.

In fact, the best thing a retailer can do to encourage impulse buying is give its customers promotions for items they regularly purchase. When we asked respondents why they bought products that were not on their shopping lists, they said it was primarily to benefit from discounts on items they often buy. But personalization is about much more than pricing. It is also about recognizing whether customers are in the store or online, remembering their preferred payment methods and receipt types, and providing them with personalized assortments.

Some of the most innovative retailers have already recognized the power of personalization. Amazon is one such example. It recently joined forces with Facebook to provide product recommendations for users based on what their friends buy.15 The personal touch has also played a large part in the runaway success of daily-deal sites like Groupon and LivingSocial (see sidebar, *Keeping Things Close to Home*).

Keeping Things Close to Home

Groupon has perfected the art of localized retailing. It persuades local merchants to offer deep discounts on products and services, sells the discounted products and services directly to consumers and pockets part of the discount. A minimum number of people must sign up for each deal before it takes effect.

The formula has proven a huge hit. Launched in November 2008, Groupon has amassed 35 million members and now serves more than 300 markets around the world, with annual sales topping US$500 million.17 Much of its popularity stems from the human element: carefully picked offers; witty, wellcrafted e-mails; and a sense of urgency, since each deal must be purchased the day it is offered.

Empower Me. Lastly, retailers have to empower their customers by making it as easy as possible for them to complete the shopping process. More than 30 percent of the people we surveyed would like to be able use mobile technologies to get help finding the customer-service desk and to place orders for goods that are out of stock, for example. More than 40 percent want to check product prices wherever they are and have promotions on items they've scanned sent to their mobile phones so they can use them when they pay for the goods. And 50 percent would happily pay via a mobile device rather than standing in a checkout line.

Consumers are equally clear about how they want to take possession of the purchases they make. Sixty percent would prefer to collect the goods themselves (even if they bought them online), while 12 percent would rather have them delivered on the same day, the following day or a date they specify. The key point is that customers want the ability to choose.

What consumers say about how they want to take possession of their goods also provides a clear pointer to the sort of retail formats they will demand in the future. Twenty-three percent of respondents told us they would like to be able to scan samples of the items they want, get an order number and retrieve the merchandise later at a designed time and place. The most innovative retailers are likely to respond by building smaller retail outlets where the emphasis is on interactivity and entertainment and keeping most of their inventory at separate sites where customers can easily collect and transport their purchases.

Removing obstacles in the consumer's path is only part of what empowering the consumer involves, though. The other part is letting customers decide how they interact with the retailer – and here there are significant regional variations. Whereas consumers in the growth markets simply want retailers to identify them through the technologies they are using and provide them with the services they require, consumers in the mature markets want to control the process by opting in and out of different services.

Smarter Consumers Demand Smarter Retailers
So what's the bottom line? Retailers that want to serve smarter consumers have to get smarter, too *(see chart on next page)*. The first step is to recognize that consumers are conducting a conversation – a conversation they control and many retailers know nothing about. Listening to this digital dialogue can help a retailer better understand its customers.

Listen and learn	Enable and execute	Empower the consumer
• Analytics to listen to consumers • Learn from consumer-controlled content • Key factors of influence and motivation	• Single view of consumer across channels • Actionable analytics around merchandising and marketing • Make the interaction personalized across channels • Change sales to service associates	• Consumer self-select interaction • Consumer choice of the interaction channel • Consumer providing suggestions

Source: IBM Institute for Business Value.

In fact, customer analytics is the most powerful weapon in a retailer's arsenal because it can uncover what is happening to individual consumers – when they experience life-changing events such as having a baby, when their children become teenagers or when they begin buying goods for their aging parents, for instance. Tagging the right data and processes and using analytics enable a retailer to create a single view of the customer, regardless of the channel through which the customer interacts; identify variations in purchasing patterns; and, ultimately, know the customer more intimately.

In the past, retailers measured their customers in terms of recency, frequency and spend. But it's now possible to look at numerous other factors that shape how customers shop, such as:

- Do they primarily shop online or in the store?

- Are they male or female?

- Are they members of the rewards program?

- Do they have a store credit card?

These are just a few of the questions that retailers can consider to better understand their customers.

The second step is to act on the insights such analyses provide – and this is no small task for complex organizations with interlocking systems and dependencies. New techniques for making information come alive can be helpful here. Data visualization, process simulation and other such tools transform numbers into information and insights that can be readily used. But it is also crucial to embed analytics into a company's daily operations by incorporating optimization logic in rules engines; altering existing workflows to allow for greater granularity in the decision-making process; and showing employees how to apply the insights analytics provide in the processes and applications they use, with case studies to illustrate the impact.[18]

With real time insights across its business, a retailer can optimize its merchandising and marketing. For example, it can rapidly identify the popularity of particular product lines, predict upcoming demand and adjust the supply chain to help ensure timely delivery of the right amount of stock to the right stores. It can also tailor its prices in line with actual and latent demand, as well as current inventory levels; develop targeted marketing programs; and take pre-emptive action to avoid shipment delays, inventory shortfalls or surpluses and other such problems.

The final step is to let consumers choose how they get the information and products they want – in other words, to make life as convenient and enjoyable as possible for them. And that requires a totally different mindset. It means recognizing that customers are not passive recipients but active participants in the shopping process, giving them the facilities they need to participate in that process and making them feel like it is a pleasure to serve them. To borrow the words of one of our survey participants, it is about providing "friendly staff who serve you with a smile, without all the huff and puff."

Responding Makes a Difference. Serving consumers – by listening to them, knowing them and empowering them – makes a real difference. For the past four years, IBM has tracked how customers' attitudes shape their shopping behavior patterns and thus how likely they are to create economic value for a company. In the course of our research, we identified a particular class of consumers – we called them "Advocates" – who are distinguished by three characteristics: they spend more with their primary retailer when it expands its assortment, they recommend the retailer to their family and friends, and they remain loyal even when rivals start offering comparable products or services.[19] This year, 37 percent of the people we surveyed are "Advocates" – up from 34 percent last year.

The percentage of consumers who spend more is even higher among those who enjoy positive social media experiences with retailers. Fifty-three percent of the consumers who follow brands on social media told us that they buy more from retailers with whom they have engaged constructively online, and 13 percent said that the increase in spending was significant. Many of these same consumers also use social media to communicate what they think. Forty-nine percent said that they interact with particular retailers or brands to submit their own opinions, while 75 percent intermittently post reviews.[20]

Creating a two-way dialogue with customers is critical in winning Advocates. It's no accident that the retailers that have enjoyed the largest gains in advocacy are those that provide their customers with personalized products and shopping experiences and use technology to empower and connect with them. They know the best way to reach their customers and that has enabled them to reduce their advertising spend or reallocate it more effectively. More importantly, it has helped them increase their revenues and capture market share from their competitors, thereby positioning them to further their lead when the economy recovers.

Conclusion

Smarter consumers are different. They are more comfortable with technology and more powerful than ever before; they make the purchasing decisions for their extended families and share their shopping experiences online. They are also more cost conscious. They want to be served – not sold to. Serving customers entails listening to them, knowing them and empowering them. That, in turn, means making efforts to understand who they are, personalize their shopping experiences and remove any roadblocks from their path so they can buy what they want when they want it. It means defining a brand promise – whether it is to offer the lowest prices, the widest choice, the most fashionable products or anything else – and delivering on that promise everywhere all the time, regardless of how consumers choose to shop.

This is a big leap for most retailers. But the smarter consumer rewards those that get smart. Past performance shows that these customers spend more money with retailers they trust. They also tell their families and friends about such retailers and spread the word on the digital grapevine. Thus, those retailers that can master the challenges of listening, learning and empowering will be well positioned to satisfy – and capitalize on – today's smarter consumer.

To learn more about this IBM Institute for Business Value study, please contact iibv@us.ibm.com.

About the Author
Melissa Schaefer is the Global Retail Research Leader within the IBM Institute for Business Value. She can be reached at maschaef@us.ibm.com.

Contributors
Jill Puleri, Vice President, Global Industry Leader – Retail, IBM Global Business Services

Robert Garf, Global Retail Strategy – Subject Matter Expert, IBM Global Business Services

Shannon Miller, Strategy and Business Development Lead, Retail Industry, IBM Global Business Services

Emmanuel Rilhac, Partner, Global Leader, Retail Center of Competence, IBM Global Business Services

Craig Silverman, Partner, BAO – Global Retail Leader, IBM Global Business Services.

References

1 Schaefer, Melissa and Laura Van Tine. "Meeting the demands of the smarter consumer." IBM Institute for Business Value. January 2010.

2 "The World in 2010, ICT Facts and Figures." International Telecommunication Union. http://www.itu.int/ITU-D/ict/material/FactsFigures2010.pdf

3 "The World in 2010, ICT Facts and Figures." International Telecommunication Union. http://www.itu.int/ITU-D/ict/ material/FactsFigures2010.pdf; "Gartner Says Worldwide Mobile Device Sales to End Users Reached 1.6 Billion Units in 2010; Smartphone Sales Grew 72 Percent in 2010." Gartner Newsroom. Gartner, Inc. http://www.gartner.com/it/page.jsp?id=1543014

4 "Facebook statistics." Facebook Web site. http://www.facebook.com/press/info.php?statistics; Murphy, David. "Twitter: On-Track for 200 Million Users by Year's End." PCMag.com. October 31, 2010. http://www.pcmag.com/article2/0,2817,2371826,00.asp

5 Elliott, Stuart. "Marketers Trade Tales About Getting to Know Facebook and Twitter." *The New York Times*. October 14, 2010. http://www.nytimes.com/2010/10/15/business/media/15adco.html; O'Leary, Noreen and Todd Wasserman. "Old Spice Campaign Smells Like a Sales Success, Too." *BrandWeek*. July 25, 2010. http://www.brandweek.com/bw/content_display/news-and-features/ direct/e3i45f1c709df0501927f56568a2acd5c7b

6 Miller, Ross. "Apple TV hits 250,000 in sales, says Steve Jobs." engadget.com. October 18, 2010. http://www.engadget.com/2010/10/18/apple-tv-hits-250-000-in-salessays-steve-jobs/Hidalgo, Dario. "Lessons learned from major bus improvements in Latin America and Asia: Modernizing Public Transportation." World Resources Institute 2010. http://pdf.wri.org/modernizing_public_transportation.pdf

8   "Transportation Statistics Annual Report, 2008." U.S. Bureau of Transportation. http://www.bts.gov/publications/transportation_statistics_annual_report/2008/pdf/entire.pdf

9 "The International Communications Market 2010." Ofcom. http://stakeholders.ofcom.org.uk/binaries/research/cmr/753567/icmr/Section_2_comparative_prici1.pdf

10 "Socio-economic profile of living conditions of older adults." *Social Panorama of Latin America (1999-2000)*. Economic Commission for Latin America and the Caribbean. http://www.eclac.org/publicaciones/xml/2/6002/lcg2068i_4.pdf

11 "The Return of the Multi-Generational Family Household." Pew Research Center. March 18, 2010. http://pewsocialtrends.org/2010/03/18/the-return-of-the-multigenerational-family-household/

12   Kohli, Martin, Harald Künemund and Jörg Lüdicke. "Family structure, proximity and contact." Chapter 4.1 of the SHARE First Results Book. January 24, 2005. http://www.eui.eu/Documents/DepartmentsCentres/SPS/ Profiles/Kohli/FamilyStructure.pdf

13   Baird, Carolyn Heller. "From social media to Social CRM: What customers want." IBM Institute for Business Value. February 2011.

14   Ibid.

15   Cutler, Kim-Mai. "Why the Facebook-Amazon.com integration is bigger than you think." VentureBeat. July 27, 2010. http://venturebeat.com/2010/07/27/facebookamazon/

16   This is based on a personal conversation with a senior manager at the company.

17   MacMillan, Douglas and Joseph Galante. "Groupon Prankster Mason Not Joking in Spurning Google." Bloomberg. December 6, 2010.

18   LaValle, Steve, Michael Hopkins, Eric Lesser, Rebecca Shockley and Nina Kruschwitz. "Analytics: The new path to value, How the smartest organizations are embedding analytics to transform insights into action." IBM Institute for Business Value. October 2010.   ftp://public.dhe.ibm.com/common/ssi/ecm/en/gbe03371usen/GBE03371USEN.PDF

19 Heffernan, Robert and Steve LaValle. "Advocacy in the customer-focused enterprise: The next generation of CRM Done Right." IBM Institute for Business Value. April 2006.

20 Baird, Carolyn Heller. "From social media to Social CRM: What customers want." IBM Institute for Business Value. February 2011

CHAPTER 2

Own the Future of Shopping

By Alison Chaltas

To be successful in today's competitive marketplace, companies need to own the future of shopping. The real focus needs to be on getting smarter about today's smart shoppers. Think of it as shopper marketing in a digital world.

Leveraging new technologies means changing the path to purchase. The old world – planning, shopping, buying, and enjoying – is fading quickly. The linchpin of change is the smartphone, which is creating the inter-connected shopper. We are now marketing in a digital age because nearly everything that we do nowadays is digital. It is nearly impossible to separate our virtual worlds from our physical world due to the growth of mobile over the last couple of years. It's different and hard to deal with, but it's exciting and challenging, too.

Studying new shopper journeys and doing a qualitative deep dive into understanding how different shoppers buy products and what their pathways to purchase really are serves two purposes: One is to understand how technology is changing shopping, and the other is to learn how we can better use technology to understand the shopper. This is the next generation of shopper research.

The Emergence of Xtreme Shoppers

There's a whole new class of consumers called Xtreme Shoppers. They represent the 40% of the population using technology to shop in new and different ways.

There are three key characteristics that are important to think about:

- *Xtreme Shoppers are wired and connected.* They have their devices in their hands or real close by at all times, and they're researching almost everything that they do.

- *Xtreme Shoppers are competitive.* A lot of this characteristic is driven out of the value equation associated with economic need. It's really shopping as a sport. Here is today's thinking: "I'm in control. I can win versus the manufacturer. I can win versus the retailer. And I'm better because of it." This is an important dynamic to consider as we start to think about how to market to these shoppers. It's not just price – it's about winning and feeling smart about things.

- *Xtreme Shoppers are much more optimistic than the general population.* One of three people believe they'll be better off in 12 months, while less than 10% of the general population thinks so. Xtreme Shoppers look at competition and technology optimistically and enthusiastically. For them, it is really about a new way of shopping to build a better life via digital/mobile means.

So shopping is going to continue to change radically because of digital/mobile. Shoppers will be relying more on tablets and smartphones. Many industry practitioners don't think this is a good thing because technology is shifting the power to the shopper, which is true. But to own the future of shopping, we need to start thinking about technology as a positive. This is the world we now live in. We need to embrace it and get on with the program.

Innovative, Proactive & In Control

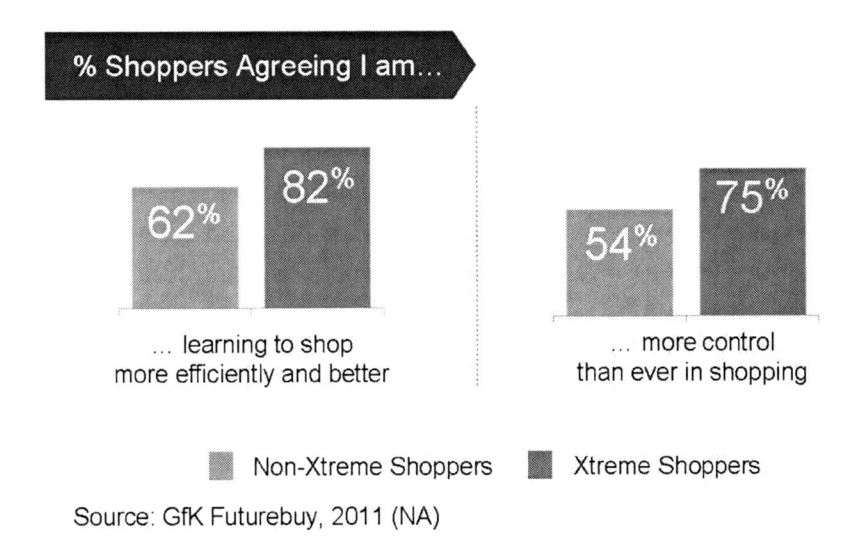

% Shoppers Agreeing I am...

62% / 82% ... learning to shop more efficiently and better

54% / 75% ... more control than ever in shopping

■ Non-Xtreme Shoppers   ■ Xtreme Shoppers

Source: GfK Futurebuy, 2011 (NA)

Moving ahead in this new order calls for figuring out how to make loyalty an important part of the relationship with shoppers, particularly online and on the go with mobility. According to GfK's 2011 *Future Buy* study, half of Xtreme Shoppers (52%) say they are less loyal to any one retailer because they need to shop around for value. This compares to 13% of non-Extremes who feel that way. Think about it. Those shoppers who are wired and competitive are going to be a lot better at shopping around than someone who is not or doesn't have that same mindset. Xtreme Shoppers are searching for deals and information online from general shopping sites, promo/deal sites, social networks, and websites hosted by retailers and manufacturers. So we need to think about developing loyalty measures versus just some immediate promotional activity.

Developing loyalty nowadays involves the mobile phone because it certainly is changing the shopper journey. Studies show that use of mobile technology is growing rapidly. Nearly half of all consumers own a smartphone, while tablet ownership almost doubled over the last holiday season. Young adults – those under 35 years of age – are driving mobile shopping. Usage tails off in the 35-49-year-old age group and then drops precipitously in the 50-plus group. Tablets follow the same pattern as smartphone use.

Studies show that four of ten smartphone owners are using their device to help them shop, while over half of tablet owners are using their tablets to do the same. And almost half of tablet owners are also using their smartphone to help them shop. So this tablet/smartphone combo is really powerful. Each plays a different role in the process, but having more mobile devices and different approaches is contributing to a different shopper journey. Overall, it's important to make the process more seamless, easy, and informative to increase conversion rates.

According to GfK's *Future Buy* study, 40% of shoppers are using their smartphone for purchasing consumer electronics, which is the most popular category for mobile use. Other categories: clothing (33%), food and beverage (27%), health and beauty (23%) home improvement (18%), lawn and garden (15%) and auto (14%). Kmart, Sears and Land's End are some of the brick-and-mortar retailers embracing mobile shopping.

There are five primary activities that are done with a mobile device:

- Value Seeking (compare price, find a coupon)

- Informational (locate store, check product availability, search for reviews or information)

- Social Media ("like" a retailer on Facebook, post a picture, rate a product or service)

- Transactional (buy products through mobile browser, buy product through app, use device to pay at cashier)

- Location (locate store, find way around store/mall, check in, use location-enabled app).

One of the worrisome activities of mobile phone use is called "showrooming." That is when consumers see a product in the store, use their phone to price-check the item, and then purchase it on their phone from another retailer or an online competitor. This is happening more often and obviously is a threat to brick-and-mortar stores. The solution requires more attention to engaging with the consumer and creating a unique experience in the store which goes beyond just price.

How do retailers differentiate? One way is to have the consumer play a role in creating the product that they want. Nike has done a good job of that with "Design Your Shoes" and Lego is succeeding with ways to design a Lego set. Many companies believe that such "co-creation" efforts are what consumers really want.

The future belongs to younger consumers and they are more open to having their experience customized. They like it when a website keeps track of their visits and then recommends things. For them, it's really a value exchange. They share information and get something in return. They are getting customization.

Shopper Marketing in a Digital Age

The same best practices that are important in traditional shopper marketing are the same for digital shopper marketing. It's not a new discipline; it's a new language, a new set of tools and communication vehicles, and a new way of thinking. But a lot of the tried-and-true practices learned over the years are still important. Here are key principles of the Shopper Marketing Manifesto to consider:

Principle #1: Know Me. Marketers need to show their appreciation of shoppers by understanding their emotional needs beyond the functional. The shopper is saying, "Know me; understand my emotions; uncover what I need, when I need it, where I am." Those kinds of dynamics are really important.

This new generation of shoppers expects one-to-one marketing.

An Opportunity for Leadership

The explosion of digital/mobile technologies and tools is the most powerful and disruptive force impacting shopper behavior. Digital/mobile is also predicted to replace many traditional tactics and to dramatically shift shopper's focus to price. Companies need to recognize this transformation. They need to understand that the stakes are high. These changes will empower shoppers to the detriment of retailers/manufacturers.

Clearly, leadership in digital/mobile is necessary to lead in Shopper Marketing.

Our national surveys indicate that the industry largely is not ready for this revolution. Only about a quarter of companies say their organization is prepared to manage the impact of digital/mobile, and that number has remained constant for a few years. The bigger picture is that the vast majority of executives agree that the industry as a whole lacks best practice perspectives to leverage digital/mobile successfully.

The leadership gap in digital shopper marketing presents an opportunity for those companies willing to make a serious commitment. Companies need to be disciplined and creative enough to determine the potential of the digital/mobile from the perspective of shopper needs instead of focusing on the technology involved.

The first step for leadership is forming a dedicated digital/mobile group within an organization. Those companies that have done so are obviously better prepared for change compared to those that have not made such a commitment. A dedicated group is a necessary and foundational step for leadership in digital/mobile.

As far as they are concerned, it's the end of many-to-one, massive television campaigns. They expect marketers to use the information that they offer up freely to get to know them and to tailor offerings to them. It's almost like we've gone back to the neighborhood store where everybody knew everybody. It's kind of using the internet as the friendly, local neighborhood.

Because of exciting GPS technology, all of our smartphones and tablets have the capacity to leverage location-based services to track and pinpoint shoppers. Opt-in GPS tracking creates endless possibilities for understanding "a day in the life" of shoppers. While it sounds like Big Brother, many shoppers are willing to allow tracking if it generates a benefit such as a customized offer. The opportunities are astonishing. For example, what's better than knowing that a shopper is not far away and a retailer can trigger a soft drink purchase on a really hot day?

Location-based services also enable companies to understand how consumers shop and track the different stores they frequent. Also, the layout of the average supermarket hasn't changed that much in a long time. So there's an opportunity to tailor that layout to how people actually shop the store as their patterns and their habits evolve.

Principle #2: Engage Me. On the surface, it's so simple to engage shoppers. First of all, they want to be engaged. So the retailer's job is to create an experience to pull shoppers in, engage their senses, and bring romance to the shopping environment. Of course, it's more difficult to get shoppers to take action and buy something. It's one thing to do that in a store, and it's more challenging in a digital/mobile environment. But it can be done, and it's not confined to very large national companies. The key is to leverage a trained staff to provide first-hand product experience to breed loyalty and community. For example, Wellness Natural Food for Pets has created a sense of community among pet owners at its website: www.wellnesspetfood.com. As well as providing product information, the site helps consumers find an online retailer or a brick-and-mortar retailer that sells its products. Information on pet grooming, boarding and training is available.

Larger companies have more resources for enhanced engagement. One of the best examples is Williams-Sonoma, which has taught the world the art of sensory, fun, consistent omni-channel conversation. Whether the consumer is in their mobile app, catalog, iPad app, or website, the experience is all the same. It is a connection between the love and the emotions that are tied to food and entertainment, and to the products that they're offering. It is all tied together, it is always consistent, and it is always relevant to the most recent holiday. Such a powerful and consistent message clearly resonates with consumers. In addition, leading consumer packaged goods companies like Coca-Cola use retail theme stores to energize shoppers and learn from them at the same time.

Principle #3: Make It Easy for Me. Mobile runs the risk of making shopping harder than it's ever been. At the same time, it can make things a lot easier. For example:

- Get me to the right place to shop

- Help me navigate the aisles and pages efficiently

- Point me to the right products

- Make things logical for me.

One of the good examples of this is Best Buy, the huge electronics retailer. These stores could be very difficult to shop, but they're not. Despite their size, the stores are organized logically, it's clear what's where, and the signs are easy to follow. Best Buy is making the shopping process simpler and they're making it much more aligned with how people actually shop the store and shop the categories. And, perhaps most appropriately, the retailer's apps echo that ease.

Principle #4: Tell It To Me Straight. Savvy shoppers want to hear the facts, and they need a consistent message. If a product or store is not going to do something, don't say it will. Shoppers will find out sooner or later. Do what you say you'll do and don't over promise.

It's all about transparency in today's digital/mobile world. Tools such as Amazon Price checker, RedLaser and Google shopper can be used by every mobile device. Shoppers can easily check prices, check the latest news, and look for any recalls. They can even check with their friends on social media. If there's something to hide, it won't stay hidden for long. On the other hand, good news travels fast in the digital/mobile world. That's an advantage to leverage.

Principle #5: Make Me Feel Smart. Making shoppers feel smart is very much linked to transparency. It's really tying together a lot of themes:

- Educate me

- Give me the best value, not necessarily the best price

- Give me ideas

- Validate and empower me

- Make me feel confident about my choices.

A good example for families that are financially challenged is the Walmart FamilyMobile plan provided by T-Mobile. The program, presented on displays in the store, gives affordable access to a mobile phone for those with lower incomes or on a fixed income. For $45 per month for the first line and $35 per month for each additional line, consumers can get unlimited talk, text and Web on T-Mobile's nationwide network. There are no annual contracts, credit checks or surprise charges. There's a phone for everyone in the family – from basic models to an Android-powered smartphone.

Principle #6: Less is More. Consumers today are busy and suffer from information overload. So cutting through the clutter is becoming extremely important. Whether it's a physical store, a virtual store, or a mobile app, it's important to lay things out simply. What messages are you trying to communicate? How do you stock what I want or make me think you're carrying what I want in new and different ways? Here are some pointers:

- Lay out the store simply so I can find my way around
- Don't bombard me with too much information
- Make it clear what messages are most important
- Carry what I want
- Give me one simple and consistent message.

The Vodafone stores in Europe used to be about selling phones. In the past, the stores were noisy and cluttered with different phones. It wasn't about leveraging the brand potential of that equipment.

The new Vodafone concept stores are very consistent across Europe and remarkably consistent with their websites and with their mobile communication. They're not selling phones per se. Yes, they sell them, but it's not what you see first. You see the décor and ambience, the service, the connectivity, and the future. You see a very different message. By providing a "less is more" environment, they are selling more phones in an aspirational way.

Mobile can simplify the world, or complicate it. An example of the former is transactional. So whether it's Walmart or some of the other retailers' online shopping lists, making it simpler and easier to buy more things is helpful. Starbucks took their purchase card and moved it onto an app. So not only can you get on your airplane flight with your iPhone, you can also use it to buy a coffee, speed up a line, and get more people through the store. Everybody's happy.

Principle #7: Keep It Fresh. How can brick-and-mortar retailers "keep it fresh" in a world where planograms are reset once a year and millions of dollars are spent for just one set of signage? Here are some pointers:

- Give shoppers new items and ideas

- Give them unique things exclusively

- Make shoppers feel a sense of discovery

- Don't over promise.

Technology can help create a sense of discovery and making sure that shoppers feel they are getting an interactive and special experience. For example, Heineken asked consumers to help co-create a new bottle for their world-class beer. Not only did they engage their customers, but the brewer also got a lot of free work that resulted in an attractive bottle. "We ended up with a brilliant design and an almost endless source of inspiration," said Mark van Iterson, Global Head of Design for Heineken.

Summary

The industry is shifting its gears from in-store marketing to truly interactive and integrated approaches that build brands at retail. The Shopper Marketing Manifesto for this new age of digital/mobile has seven principles that need to be followed to get on the road to success. By understanding Xtreme Shoppers and those who will to join their ranks, companies can own the future of shopping.

Alison Chaltas is an Executive Vice President with GfK Shopper & Retail Strategy, part of the GfK Group. She has spent more than 20 years at the forefront of shopper marketing and category management, building brands through creative insight-based retail solutions. More information: www.gfkamerica.com.

SECTION TWO
LOYALTY

CHAPTER 3

The Retail Arms Race

By Gary Hawkins

It can be argued that today's most important retail battle started on May 12, 2003. It is a battle being waged largely out of view. Many in the industry – particularly independent and mid-market retailers, and even some wholesalers – are largely unaware. Like all battles, it is fueled by materiel and fought with weapons used by soldiers. There are too many casualties.

The material fueling this war is data – lots of it. Transaction data and shopper data are central, but there are other sources. Demographic data, in-store video monitoring, mobile-based location data, real-time social media feeds, third-party data appends, weather, and many other types of data are all playing a role.

The soldiers in this retail battle are not front-line store personnel; no, they are high-priced data analysts, hidden away at headquarters. High-priced mathematical talent has found its way to the retail industry, drawn by retail's vast storehouses of virgin data.

The weapons: Shopper Insights mined from the ever-expanding data streams that provide opportunity for retailers and brands to improve marketing, merchandising, operations, and many other activities throughout the enterprise. It is scale that drives the value creation powered by these insights.

The casualties: Independent and mid-market retailers. The number of independent retailers in the U.S. has fallen to an estimated third of the number that existed thirty or forty years ago; at the same time, the number of chain stores has dramatically increased, driven by the largest retailers. The independent and mid-market retail sectors that have long provided rich diversity are fast becoming cannon fodder.

So what happened on May 12, 2003 to trigger this battle? Kroger announced its partnership with dunnhumby, triggering a shopper data arms race.

Kroger is the nation's largest traditional (if there is such a thing anymore) supermarket chain; dunnhumby is the analytics consultancy that helped power Tesco to its leading position in the UK market through mining the shopper data gathered through Tesco's loyalty program.

One does not have to look far to see the partnership's rewards. Kroger recently announced 34 consecutive quarters of same-store sales growth. David Dillon, Kroger's CEO, proclaimed dunnhumby as his "secret weapon" in cutthroat battles with rival supermarket retailers.

In a very real sense, Kroger – through its partnership with dunnhumby – has effectively *weaponized* shopper data in several ways:

Personalized Marketing. Kroger-dunnhumby leverage the data to provide highly targeted promotions to its customers, driving basket size, shopping visits, and retention over time. Providing savings on relevant products to each customer household is powerful. Kroger has more than doubled the number of premium loyal shoppers.

Merchandising. Using the shopper segmentations and insights created by dunnhumby, Kroger drives this knowledge down to a store level, using it to tune each store's assortment and inventory to the store's shopper base. Such an approach delivers improved return on inventory and lowers out of stocks as brand assortment is better planned relative to the store's shoppers.

Direct Monetization. Kroger charges brand manufacturers millions of dollars each year to access shopper insights and analytics provided by dunnhumby; the suppliers in turn are expected to use the information to help Kroger improve its business. "The partnership is generating millions in revenue by selling Kroger's shopper data to consumer goods giants ... 60 clients in all, 40% of which are Fortune 500 firms" states Matthew Boyle of CNN Money in a report on Kroger. The retailer is realizing over $100 million annually in incremental revenue from these efforts.

Like subcutaneous hemorrhaging, the devastation caused to competing retailers by this battle is often out of sight. Kroger has more than doubled the number of premium loyal shoppers. It is converting secondary, lower-value shoppers into loyal, high-value shoppers. This growth in revenue is coming at the expense of other competing retailers, often independents and mid-market stores. Increasingly, Kroger is becoming the primary shopping destination for a growing number of consumers, while the independent / mid-market retailers it competes with are relegated to the role of glorified convenience stores. Their highest value shoppers are being siphoned away. And without shopper data, a retailer may not even realize what is happening. Beneath the surface, margins deteriorate as the store becomes used for secondary shopping trips, depriving the retailer of a healthy mix of products in the basket.

The success of the Kroger-dunnhumby partnership has changed the nature of competition in the retail industry. Kroger is effectively forcing the creation of a new battlefield that independent and mid-market retailers are largely precluded from: Shopper Marketing.

Broadly used, the term Shopper Marketing can mean many things to many people. For our purposes, we shall define it as *using strategic insights on shopper behavior to influence the shopper on the path to purchase.* Shopper Marketing initiatives are often paid for by manufacturers out of dedicated, retailer-specific budgets and initiatives are built on collaboration between the brand and retailer.

But focusing the discussion on only Shopper Marketing constrains the view to what is happening at a broader level where there are more far-reaching ominous trends at work. While there has always been a performance gap between the largest and smallest retailers created by economies of scale, Shopper Marketing and related trends are rapidly – and dangerously – widening this gap at an increasing pace.

We see several major trends that pose an increasing challenge to the independent and mid-market retail sector:

Big Data. Though a term at risk of becoming over-used, Big Data remains a threat. Big Data is the focus of the largest retailers and manufacturers, supported by leading technology companies and consulting firms. It can be defined by the three "V"s: volume, variety, and velocity: Massive Volumes of all types of data, growing from petabytes to zettabytes; Variety driven by increasing ability to measure everything (transaction data, shopper data, twitter feeds, mobile-location data, and many others); and all this moving at high Velocity, monitoring all these data streams in real time. Big Data is typically unstructured, requiring new technologies like Hadoop and NoSQL to analyze, and even new approaches like Data Science to manage it. McKinsey, in a May 2011 report, projects that a retailer using Big Data to the fullest could increase its operating margin by more than 60%. Independent and mid-market retailers simply have no access to this level of sophisticated and comprehensive capability.

Shopper Data. Customer-identified transaction data, typically gathered through loyalty programs, is creating a digital divide in the retail industry; those retailers that are customer intelligent and those that are not. But even shopper data is not enough today; marketers are procuring additional data from third-party providers on demographic lifestyle attributes, media preferences, and additional consumer segmentations. Even retailers without loyalty are collecting vast amounts of data, creating "virtual" shopper profiles by linking basket data via payment information, appending demographic and lifestyle attribute data at a store level based on the surrounding customer base, and using panel data.

Shopper Insights. Legions of data analysts at the largest retailers and consulting firms are mining the massive data stores to generate powerful shopper insights. Target's dedicated analyst group became so good at this that they were able to determine that a teen girl was pregnant before her father knew (reported in a story "How Companies Learn Your Secrets" in the *New York Times* on February 16, 2012). The article illustrates the power of shopper insights.

As many in retail understand, changing a consumer's shopping habits is challenging as store and product routines become ingrained. But analysts have discovered that there are times in life that create flux and an opportunity to reset ingrained habits. One of those times is around the birth of a child; parents are exhausted and store and brand loyalty are up for grabs. Target, understanding this opportunity, sought to discover "new parents" before birth records were published, sending competing retailers in pursuit of the prospects. Target's analytics team was able to do this by discovering small changes in purchase behavior that indicated a shopper was pregnant – even what trimester she was in – enabling marketers to reach out with special offers at the opportune moment. And this is just one small example.

How can even the most successful mid-market retailers bring such resources to bear, let alone smaller retailers?

In-Store Research. Major consumer packaged goods (CPG) manufacturers, often in partnership with the largest retailers, are heavily focused on understanding true in-store shopper behavior using a variety of research techniques. This focus is driven by massive growth of shopper marketing initiatives seeking to influence the shopper's decision on the path to purchase. The 2012 Shopper Engagement Study just released by the Point of Purchase Advertising International (POPAI) discovered that a whopping 76% of purchase decisions are made in the store – an all-time high. Little wonder brand marketers and retailers attach such importance to understanding and influencing in-store behavior.

Many large brand manufacturers have created dedicated research facilities to focus on shopper understanding. Some facilities, like Procter & Gamble's BRIC, actually recreate a store environment where shoppers can be observed; others create virtual reality environments where test subjects shop. The science of neuromarketing, which has gained steadily the past few years, has researchers monitoring how the brain reacts to different stimuli like package design, advertising messages, and so on.

If all that were not enough, the research has moved in-store. Researchers are bringing sophisticated tools to bear: video and mobile monitoring and analytics, eye-tracking technology, even smart shelves that record when a product is picked up, are providing powerful new insights to actual shopper behavior in the store. Many of the largest retailers, often working together with their key brand partners, are dedicating stores within their banners to serve as live learning labs to test the efficacy of display designs, merchandising concepts, and product locations in the store. In-store merchandising is rapidly becoming a science driven by an explosion of data. Independent retailers are being left behind without access to learning and best-practices.

Technology. While the largest retailers have increasingly leveraged technology to drive operating efficiencies and reduce product logistics costs, the performance gap with smaller competitors is expanding quickly as sophisticated technologies are brought to bear on optimization of pricing, promotions, and product assortment. Even this wave appears small compared to the tsunami coming next: sophisticated real-time marketing personalization capabilities.

Walmart recently acquired a social media company, Kosmix, creating @WalmartLabs, with the intention of creating an R&D unit to define the future of commerce by merging social, mobile, and retail. The group created a gift recommendation app called ShopyCat that suggests gifts for friends based on their Facebook profile. WalmartLabs is also focused on in-store navigation using mobile: apps that offer customers information and the location of recommended items, or prompts for items of interest that are nearby. Here is an example of where they're going: Imagine walking by the Walmart sports section and having your mobile phone remind you of a friend's upcoming birthday and interest in fishing, helping you make a relevant purchase while saving time and hassle.

Marketing Funds. Like many battles, this one is ultimately over riches; in this case, it's industry marketing funds. A recently released study by the Grocery Manufacturers Association (GMA) estimates annual industry spending on Shopper Marketing at over $50 billion, and growing. A disproportionate amount of this vast sum is going to the largest retailers where manufacturers can collaborate on specific initiatives leveraging shopper insights, and target specific promotions to specific shoppers to achieve marketing goals.

This growth in Shopper Marketing budgets comes as manufacturers are simultaneously reducing spending on traditional trade promotion. A 2012 trade promotion study by Kantar Retail shows manufacturer spending on trade promotion, measured as a percentage of gross sales, at the lowest level since 1999. But even this does not tell the whole story; it is the changing *mix* of manufacturer marketing expenditures that shows what is occurring. Trade promotion accounted for 44% of total marketing expenditures by manufacturers in 2011, lower than any other year in the past decade. This decrease is driven by a corresponding increase in Shopper Marketing expenditures.

Here is the most worrisome part of this trend: a disproportionate share of these vast marketing funds is directed to the largest retailers. Those are the retailers possessing powerful shopper insights and the technology and willingness to direct specific promotions at specific shoppers to drive business in collaboration with the key brands.

This story is just beginning. The Kantar Retail report goes on to say "manufacturers anticipate that changes in the next three years will revolve around continued trade integration with Shopper Marketing to maximize value in the face of continued margin demands.

Manufacturers, in particular, expect to allocate trade funds more strategically in the future, as they shift to a "pay for performance" approach and more closely measure program and retailer performance."

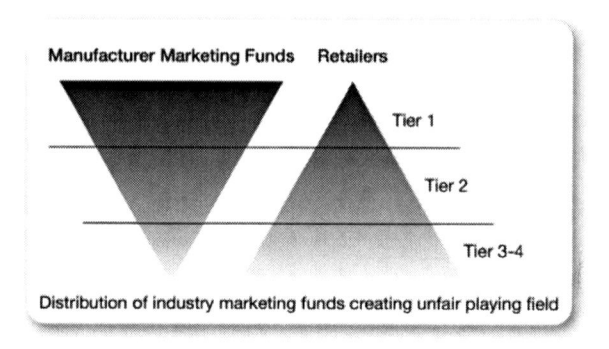

```
Manufacturer Marketing Funds    Retailers
                                    Tier 1
                                    Tier 2
                                    Tier 3-4
Distribution of industry marketing funds creating unfair playing field
```

Shopper insights are driving increasing marketing personalization; that is, those personalized promotions funded by incremental Shopper Marketing funds paid for by the large brand manufacturers.

Personalized marketing is the advanced weaponry of this retail battle; it is retail's version of laser-guided smart bombs. From Kroger's mailing of highly relevant promotions to its shoppers to Safeway's Just for U program delivering savings on relevant products to its loyalty program members, this battle is well underway. Ahold is pursuing a similar strategy; CVS holds the leading position in the drug store channel, delivering personalized, relevant promotions to members of its ExtraCare program. Personalized marketing has proven its ability to drive increased basket size, increased shopping trips, and increased shopper retention over time. And if you're a top-tier retailer, you get all these benefits paid for by manufacturers funding those personalized promotions to your shoppers.

This personalization, being done on a massive scale, is far beyond what many retailers envision; these initiatives are being driven by high-powered analysts (dunnhumby has over 120 analysts focused on Kroger alone!) and, increasingly, next-generation web-based personalization engines are being put to the task. The most advanced of these technologies are able to watch in real time what shoppers are looking at or searching for when online via their PC or mobile, and are able to serve up relevant promotions, content, cooking videos, recipes, and more. Another solution provider works with Tier 1 retailers using their loyalty data to link via cookies to the same shopper online and be able to present web advertising based on purchases in the physical store. The future of retail is about delivering the right promotion to the right shopper at the right time and in the right place.

Retailers continuing to go to market with "no loyalty card required" or "same prices for everyone" are destined to become a quaint anachronism as shoppers of all ages increasingly expect to be catered to through savings on products and services relevant to them.

The cumulative force of these trends is laying siege to a wide swath of the supermarket sector. Independent and mid-market retailers are increasingly unable to keep pace because they lack in-store best practices, sophisticated technologies, and powerful shopper insights. The growth of Shopper Marketing budgets and the distribution of a disproportionate share of those funds to the largest retailers are slamming independent / mid-market retailers from both sides. There is little ability to access these incremental promotion funds to drive shopper spending while reduction in historical trade promotion decreases competitiveness.

Wholesalers, who many independent and mid-market retailers rely on to procure manufacturer deals, are able to offer little help. Wholesalers have no shopper data and do not possess the sophisticated tools and analysts required to power shopper insights used to gain access to new manufacturer marketing funds, let alone any digital channels to the shopper. The problem is made worse by the fact that wholesalers – already challenged by retailer compliance with regard to regular trade promotion – lack the scale requisite to fight this new battle.

This growing retail industry schism between the very largest retailers and everyone else is only going to get worse. The same Kantar Retail report strongly calls out that the future success model will involve deeper and more extensive collaboration between the retailer and brand, with focus on clear objectives and performance accountability. This is much easier for the brands to accomplish when deploying entire teams of people against a Kroger or Target or Walmart; it is much harder – some could say impossible – when having to interact with hundreds or thousands of mid-market and independent retailers. These companies are very much at risk of losing their most important historical advantage – connecting with the customer – when the largest retailers have more knowledge of that shopper's preferences and caters to her.

In 1936, the Robinson-Patman Act was signed into law to address anti-competitive practices – specifically price discrimination – and require sellers to offer the same price terms to all customers. As manufacturer marketing funds have grown increasingly important to driving retailer sales, the question the industry faces today is clear: Is a similar environment taking hold in the form of growing inequality and unfair playing field? The forces at work in retail threaten the very diversity of the industry.

Admittedly, this is a rather bleak view of the future for independent and mid-market retailers. Is all lost? Not necessarily, but time is running short. Before one can rectify a problem, one has to first acknowledge that there is a problem. That's one of the goals of this chapter: To identify these key trends and the impact they are having. When one is head-down in day-to-day operations as most retailers are, it is very difficult to step back and view macro industry events.

Hopefully this chapter can help start a discussion. The continued shrinking of the independent and mid-market segments threatens the retail industry, depriving it of a diversity that drives innovation and new, exciting developments. The supermarket industry in particular is having enough challenges; a recent article in the Wall Street Journal ("What's Wrong with America's Supermarkets," July 12, 2012) calls out a study by UBS showing that supermarkets share of US grocery sales fell to 51%, down from 66% in 2000. A major cause: big box discounters like Walmart and Target using food to attract shoppers to their stores.

This chapter can serve as a call to arms for the independent and mid-market retail sector. Referred to as "independent" for a reason, retailers in this segment of the industry must understand the forces at work and make informed decisions regarding the future of their businesses. While technology advancements are affording big retail clear advantages, other advancements are benefitting smaller businesses as increasingly sophisticated solutions are made available via the cloud. But it is incumbent upon the retailer to be open to new initiatives and devote the resources to take advantage of new capabilities.

Gary Hawkins is CEO of Hawkins Strategic, a consultancy focused on gathering, understanding and using detailed customer data in retail. He works with retailers, manufacturers and technology companies worldwide to optimize customer loyalty data and value. For more info: www.hawkinsstrategic.com.

CHAPTER 4

Change Management by the Numbers: Devising a Strategy That Produces Results and Advocates

By Brian Ross

Corporate history is rich with tales that heed the need for proper change management to drive value from new strategies, plans and tactics – so much so that over the past few years we have seen a tremendous rise in the prevalence and credibility of the discipline. This should not be surprising, given that in retail all value lies in execution.

Shopper centricity, for instance, has been one of most significant trends driving retail growth globally. While there has been much success, there are an equal number of challenges that cause many organizations to rethink their consumer strategies. This presents the risk of not realizing the significant strategic and financial benefits available.

Simply put, we know that shopper centricity works when properly engineered. It delivers a significant and sustained increase in sales and profits, and provides a foundation for long-term competitive differentiation.

Of course, it is not always done right, but at least we can learn from our experiences. From our research and from experiences working with retailers and manufacturers, we know that in almost every case the root cause of success or failure in a shopper-centric plan lies as much or more in effective implementation and change management as in the quality of shopper insights and the strategic recommendations thereof.

To put it another way, shopper-centric strategies don't fail because we don't know what to do; they fail because we sometimes don't know how to execute, or lack the right support to execute.

Whether you are a retailer or manufacturer, adopting this shopper-committed approach – be it through pricing, assortment, store design or marketing – takes more than data, marketing algorithms and a new set of rules. Rather, it requires a comprehensive set of tools, people and processes aligned behind the plan. And, importantly, it requires company-wide support, all the way to the top executives.

This change can be daunting, but it doesn't have to be. We rely on a tried-and-true process that has worked for many clients. But it takes a measured approach to planning and prioritizing, a review of product performance and a gauge of management's commitment across ranks.

First, let me define what change management means within our organization and how we apply this to a shopper-committed strategy. We believe that successful change management stands on three pillars:

1. **Create a Shopper-Centric Approach to Decisions and Tactics:** A shopper-centric strategy requires changes to how we make decisions today. This means integrating new shopper data and insights into our existing processes to determine what our growth measures should be. Do you want higher volume or bigger planned trips among your best shoppers?

2. **Develop New Processes and Tools:** We should incorporate our newly determined approach into the process, make necessary

changes, and support these changes with the appropriate data and tools.

3. **Align the Organization:** We have all heard the axiom that what gets measured gets managed. For true change to be implemented, the organization will want to measure and manage its processes and teams in accordance with its new shopper-centric approach. For example, if a category manager is evaluated on traditional metrics of sales, share and margin, and not on new shopper metrics, it's unlikely that his or her behavior will change in the long run. To support the change to shopper centricity, all of the organizational structure, roles and policies – right down to how people are measured and evaluated – must change.

These basic pillars establish the platform upon which an organization can build its change management plan toward achieving shopper centricity. But the real question is how do you get there? The following five steps will walk us through the process of building and launching a foolproof pilot plan.

Teamwork and Testing: Five Steps to the Pilot

1. Develop a shopper-centric strategy. The first step is to establish an enterprise-wide strategy through comprehensive segmentation and insights. The key hallmarks of an effective action plan include: identifying priority shoppers; understanding their needs; assessing the strengths of your current performance; and prioritizing opportunities to increase sales, profits and loyalty. It is vital to have senior and cross-functional organizational commitment at this point so that all marketing and merchandising decisions are made to meet the needs of priority shoppers. This will ensure that everyone's expectations are in alignment, from the corner office to the aisle. It also positions the organization to make important personnel decisions, such as appointing a core Change Team to identify the key roles of staff members, and establishing accountability and reporting structures.

2. Prioritize your shopper-centric tactics. Where to start? One of the most significant decisions in effective change management involves picking the best-suited initiatives to drive the greatest value. At Precima, we have successfully used a formula that cross-references the expected and potential value of the chosen tactic with the organization's ability to execute the plan. The result is the following opportunity matrix:

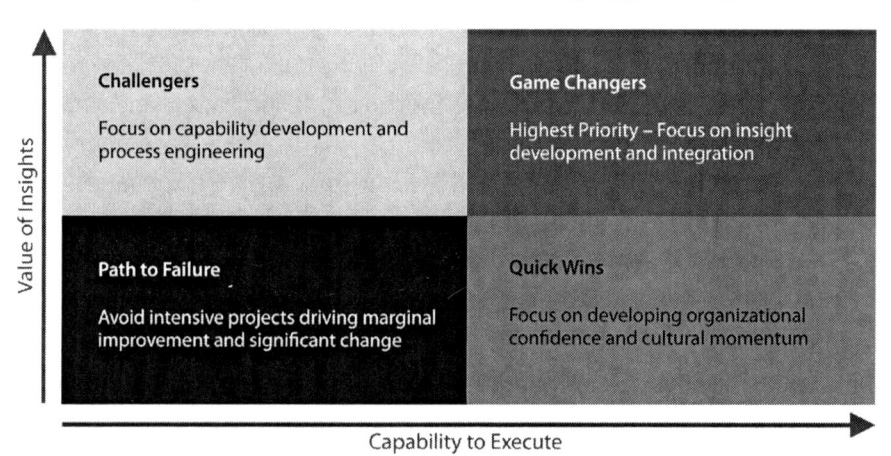

	Low Capability to Execute	High Capability to Execute
High Value of Insights	**Challengers** — Focus on capability development and process engineering	**Game Changers** — Highest Priority – Focus on insight development and integration
Low Value of Insights	**Path to Failure** — Avoid intensive projects driving marginal improvement and significant change	**Quick Wins** — Focus on developing organizational confidence and cultural momentum

The first opportunity quadrant includes the **Game Changers.** These are the shopper-centric strategies and tactics that are applied to categories and stores that possess both significant insight value and high execution ability—a sure-fire success. Game Changers should be the highest priority for building a shopper-centric organization.

Next are the **Quick Wins**, which have lower insight value but high execution ability among managers and resources. These initiatives help to build organizational confidence. The Quick Wins are followed by the **Challengers**, those opportunities that have high insight value but are resistant to change. The focus here should be to develop capabilities to improve executional capability.

Lastly there are the **Path to Failure** initiatives. These approaches need to be identified early and avoided at all costs as they will consume time and effort while delivering no value to the organization. What is worse, these failures will create naysayers and skeptics in the organization and derail the journey to shopper centricity.

3. Define your shopper-centric approach. Once your tactics are prioritized based on your goals and the change opportunity matrix, you'll want to take your strategy to the next level through analysis and insights that guide your shopper-centric approach toward the most effective organizational decisions and tactics. These insights lead to the kinds of decisions, recommendations and strategies that will drive increased sales and profits and long-term shopper loyalty. It is essential to develop a shopper-focused approach to the selected strategy and build the analytics, recommendations and test plan based on how the strategy will be executed. This will require a different decision-making process, one that examines traditional tactics through the new scope of the priority shopper.

4. Develop the Pilot plan. Here is where we are going to apply the elbow grease. In partnership with the newly formed Change Team, an Execution Team will be required as the key agents for change, developing the plan, defining the Pilot parameters and leading the execution and measurement. Together, the Change Team and Pilot Execution Team are tasked with devising the plan that will build the proof points for the Rollout.

It is vital to work on a manageable scale and to keep the lines of communication open throughout this process. The plan will only succeed with the right-sized tools and with support teams that are accountable and exceptionally open to change, as both characteristics will be necessary to achieve your desired goals. And it is absolutely critical to provide the right level of support. In fact, it won't hurt to over-invest in your support levels since the team members will be learning as they go along, and often while also seeing to their regular work duties. For these reasons, you'll want to have your key benchmarks and definitions for success in place. Ask yourself: What are we trying to achieve, and what does success look like? For many of us, there is a tendency to rush this step or to determine it during the execution phase, but resist the urge. It is often a recipe for failure.

5. Execute, measure and build the case for a Rollout. Once the plan has been developed, it's time to execute the Pilot in those test categories and markets. Again, the most effective approach involves developing a series of tests, not just one, and continuously measuring the effects of each. This includes not only the strategic and financial benefits derived from the measurement plan, but also your own discernment of process efficiency, effectiveness, best practices and opportunities to enhance program benefits. The financial results build the case for change, while the organizational results build the proof that change can be implemented. The Pilot Execution Team and Change Team are both tasked with building the business case for change, based on the test results, the effect the proposed change will have on the organization, and then the recommendations for execution—the what, the why and the how and when.

Once you've completed these five steps, you will arrive at a moment of truth for the Pilot-to-Rollout process. That is when it transforms your Pilot Execution Team into Agents for Change and Ambassadors for Shopper Centricity. In fact, that is one of the key goals and rewards of the Pilot approach: to create credible and vocal advocates for change who drive organizational buy-in, from the executive suite to the front lines.

These advocates will become the primary internal sales teams for change, the "help desk" for questions and communications. They also will often become the "continual" Pilot team, regularly driving new enhancements and extensions to shopper-centric approaches, while also serving as the ultimate how-to guides. In other words, they should become the most trusted peers among the company's rank and file, speaking the language of the rest of the teams. They can say to the organization, "Here's how we did this, here's how it worked, and you can do it, too."

But a misstep anywhere along the Pilot process can cost your teams some of that credibility. For this reason, I cannot emphasize enough the importance of implementing change in a measured approach and never underestimating the amount of work it takes to do so properly. It is essential to adhere to the tenets of successful change management and to recognize that change management doesn't end with the plan, but rather that it starts there. Communications, management, measuring, refining and documenting are the key to transforming the pilot into a Rollout.

After all, there is nothing worse than making the wrong first impression – with your customers or with your employees. The key to engineering any successful structure, whether it is an office tower or a shift in company-wide operations, is a solid foundation. In the case of change management as a best practice, that foundation is a living entity, and its greatest strength is the ability to continuously test, measure, refine and communicate change throughout the process.

The Proof Is in the Footing: Rolling out for Change

Once your foundation is in place, it is time to move forward and roll out the concept throughout the company with your team of ambassadors driving the change. But keep in mind that while the Pilot built the business case for organizational Rollout, this by no means guarantees success of expansion. The Change and Pilot Execution Teams will run into significant new challenges upon organization Rollout, including the effects of scale, internal "pockets of resistance," adjusting for phasing and sequencing shifts, and learning how to communicate and coordinate change on an exponentially larger scale, to name a few.

But by first following those five steps to an effective Pilot, we've established the strongest foundation for future success, driven by an organization-wide commitment to the planned change, proven results, best practices and a dedicated team to act as ambassadors and implementers of the change. Advancing to organizational implementation is the most vigorous phase of change management, building on the foundation we established in the Pilot phase.

We rely on an intensive, three-step process.

Step One: Build and Communicate the Plan for a Rollout

The most important step to planning change is managing and communicating change. Leverage what was learned from the Pilot since its results justified the change, whether the goal was to increase sales or profit, improve operational efficiency, or enhance service.

There are three elements necessary to doing this:

- *You must have commitment.* A company really needs strong, unwavering executive-level support that extends to those most impacted by the change. People tend to resist change naturally, so associates will need to see and believe the reason for the change, understand how and why it impacts them, and have confidence that the change is being properly supported by the organization. Top executives, along with the Change and Pilot Execution Teams, need to consistently champion the change, assign adequate support, share accountability and— never forget—communicate, communicate and communicate the plans.

- *It must be comprehensive.* Here is where the full-time Change Team will prove its mettle. It will be up to these members to be sure workers are provided with the tools and processes to roll out the change and foresee potential hurdles. It is paramount that all elements—people, tools and processes—are aligned and integrated as the change is expanded. This phase is often called the "plan-the-plan" phase, and it is the critical step for success as it is next to impossible to build the plan once you've started a Rollout.

The benefit of a Pilot-to-Rollout strategy is that you can build key findings on a limited scale prior to a broad organizational Rollout. But these incremental shifts do come with a risk, and that is the Pilot strategy might not expose the Change or Pilot Execution Teams to all of the shifts that are required organization wide. One remedy is to build the plan with some wiggle room so that you can expand the team when needed to ensure that the proper people, tools and process are in place to enable the change. Remember, while the Pilot can operate with "duct-tape and bailing wire," that approach is not sustainable at the scale of organizational Rollout.

- ***You have to communicate from the boardroom to the aisle.*** Effective, frequent and open communication is at the heart of successful change management. At every step of the way, the Rollout's plan and progress must be made clear across the organization, and the message should be coming from the every level, not just the executive level. This includes explaining at the onset why the change is needed, what the plan entails, what to expect of every employee, and detailed updates of results and refinements. For change to work, people have to understand and believe in the reasons the change as well as its expected deliverables.

Step Two: Execute the Rollout

Now that you've built and communicated the plan, it's time to execute the plan throughout the organization in a phased manner. This is the moment of *proof*, when you will determine the value of the change to the organization and whether your strategy is a success or failure.

But even if it is a success, your work is not done. Far too often, we tend to believe that once the plan is built and communicated, our work is done and that the execution of our strategy can run its course on autopilot. That is not a good idea. The reality is that execution and implementation will only hum along if we continue to adhere to these three basic practices:

- ***Continue to communicate.*** Yes, it bears saying twice. But this time, let's emphasize the listening part. What the Change Team hears may be as important as its own message. Remember, the plan is not rigid; it is not pure science; it's a social science. Once the implementation begins and other employees get involved, your teams will likely experience some resistance, challenges and surprises. But they also will hear some new ideas that warrant careful consideration, which leads to the next point.

- ***Be flexible.*** While the implementation plan is built at a very detailed level, this shouldn't lead to a rigid adherence to the plan or lack of willingness to change. Rather, it is a good idea to leave room for adjustments as the plan is implemented. Not every element could have been anticipated and not every step of the Pilot will be replicable on a large scale. Use the plan as a guide and the Rollout as an opportunity to learn more about your capabilities, and to optimize.

- *Manage Change, Part II.* The Change Team should continue to meet with senior executives to routinely identify pending problems and cut a clear path to resolving them. To succeed, the Rollout needs to be managed meticulously and rapidly; don't wait until potential land mines explode into major problems.

Step Three: Optimize Change

The best way to wrap up a change management Rollout is to never wrap it up. There is no end date to effective change; it is an ongoing process and it is essential to have measurements installed to determine what is working and what is not.

The following straight-forward elements will help ensure your ongoing success:

- *Measure continuously.* Track the impact of the change as it relates to the original goals, whether they are to improve sales, increase shopper visits or reduce inventory expenses.

- *Monitor your staff's performance.* Change management often requires new or different responsibilities for workers, and sometimes it can take months for someone to reach his or her highest productivity in that role. Be sure to ask for new ideas and feedback.

- *Keep the band together.* In this case, that's the Change Team that so successfully led the process. These advocates have the institutional knowledge to understand what did work and the wisdom to recognize what did not. They are the Geiger counters that will detect unwanted shifts and ensure the change is going smoothly.

In other words, the secret to effective change is to continually foster change. It is a cycle of proficiency, steadily gauging results and then carefully shifting the strategy to enhance your performance. And then it continues to the next-highest level. The Pilot positions your gears, and the Rollout places the machine into motion, keeping it well oiled and at optimum performance. Your new ideas for running it more efficiently simply fuel the engine.

There's no question that change can be onerous, and doing it right takes a rigorously disciplined process. But if the time is invested wisely in the right Pilot and Rollout processes, then the results will pay off in long-term dividends. Shopper-centric strategies, when successfully implemented, have proven to increase sales and profits by 5% to 10% annually. But just as important, a shopper-committed focus is a differentiator – a strategy that cannot be easily duplicated because you rely on your own data and talent to align your brand to the needs of your best customers, which in turn enhances long-term loyalty.

Yes, it takes a bit of tinkering. But once we heed the need, this kind of engineered change is what we all should strive for.

Brian Ross is president of Precima, a part of the LoyaltyOne group of companies. Precima applies the insights gained from advanced shopper analytics to help retailers and manufacturers increase sales, boost profits and build long-term loyalty. For more information, visit www.precima.com or email bross@precima.com.

CHAPTER 5

Actionable Shopper Segmentation

By Megan Margraff

Actionable shopper segmentation aims to identify groups of shoppers with similar needs and behaviors so that each group's distinct needs can be met. The most effective shopper segmentation strikes the right balance between two competing objectives: on the one hand, to recognize that not all shoppers are alike, and that retailers will require differentiated strategies to meet the spectrum of shopper needs; and on the other hand, to recognize that many of these differentiated strategies will be implemented in an environment in which key variables, such as price, assortment and merchandising, cannot be tailored to the individual household. Simply put, actionable segmentation balances the need for complex approaches with the need for a simple framework for execution.

Why does shopper segmentation matter? For the grocery retailer, shopper segmentation should be the foundation for how decisions that affect the shopping experience are made. The most customer-centric retailers leverage shopper segmentation across departments – from insights to marketing to merchandising and category management. For consumer packaged goods (CPG) manufacturers, shopper segmentation provides a unique means of determining where the brand and the shopper intersect. As retailers commit to using shopper segmentation as a foundation for better decision-making company-wide, manufacturers who embrace the opportunity to partner with retailers to tailor plans based on shopper insights will be rewarded. Those who fail to do so will become increasingly irrelevant.

Let's look at various approaches to shopper segmentation in terms of the techniques used and the shopper-specific inputs considered. Next, let's discuss best practices in terms of how shopper segmentation is implemented and leveraged organizationally by grocery retailers and their manufacturer partners.

The Basics

In the earliest stages of shopper segmentation, the question for most grocery retailers is simply, "How do I recognize and reward my best shoppers?" To answer this question, retailers with customer data often begin by segmenting shoppers based on a handful of variables related to the value of the shopper to the retailer. Two common approaches to segment shoppers simply based on their value are deciling and quadrant-based segmentation.

Deciling involves ranking households on spending (typically annual or quarterly net spending) and breaking them into deciles, each of which represents 10% of shoppers. Deciling allows retailers to understand what percent of total sales derive from their top 30% of shoppers (typically upward of 70%), but not much else. Even at its best, decile-based segmentation provides no meaningful clues as to why shoppers are in upper or lower deciles and what different shoppers need to grow their spending.

Basic quadrant-based segmentation represents a step up from deciling, and involves using "hard" breaks based on two and sometimes three variables to place shoppers into segments; for instance, Platinum, Gold and Bronze segments using a combination of recency, frequency and spending. Basic quadrant-based segmentation is more useful than deciling as an ongoing tracking tool for retailers, but quadrant-based segmentation suffers from the same blind spots as deciling; that is, the basic variables used typically don't provide meaningful information on how to meet different shoppers' needs.

Beyond the Basics

As retailers become more shopper-centric, the nature of the questions the shopper segmentation is intended to address evolves, focusing more on the shoppers' needs and how the retailer can meet those needs than the value of the shopper to the retailer. Retailers want to understand how shoppers are engaging with their stores. They are asking questions like:

- How broadly are my shoppers shopping across the store?

- Do they have key department or category voids?

- Are they sensitive to price or promotion?

- Do they shop the front page?

- What are their typical shopping patterns?

- How long and how consistently have they been shopping with me?

Retailers also want to know how shoppers place their stores in the context of competitive options:

- How loyal are my shoppers to me?

- What share of my shoppers' grocery spending am I capturing?

- How do they rate my stores relative to my competitors?

Retailers also want to know more about their shoppers' lives to serve their needs better:

- How many of my shoppers have kids?

- Which prefer premium or value products?

- Which of my shoppers need quick meal solutions or healthier options?

Retailers on the cusp of developing more advanced shopper segmentation have several important questions to consider:

- What data do I use?

- Where do I begin?

- What segmentation technique should I use?

What Data to Use? Where to Begin?

Let's take the first two questions together: What data do I use and where do I begin? Retailers have many data options when considering how to segment shoppers. Customer data can provide a wealth of calculated metrics capturing key shopping behaviors for each household. When considering the use of customer data, retailers should go beyond the top two or three metrics typically used in recency-frequency-spend segmentations and consider trip size vs. frequency, deal and price sensitivity, consistency of shopping, breadth of shopping across departments and categories, shopping patterns by week, day and time of day, and skews to key departments and categories. Basic demographic data is available from many suppliers, and the cost of appending this data to customer lists has come down enough in recent years that it is within reach of even smaller retailers. In addition, many modeled share of wallet and lifestyle household-level data can also be purchased. And, of course, surveys to a sample of shoppers can provide retailers a glimpse of shopper attitudes, preferences, and satisfaction ratings for their own stores and key competitors.

The key question is not what data should I use (all of the above is a good answer) but rather, with what data do I begin the segmentation process? Data elements should be separated into *driver data* (variables that will be used as active inputs to separate shoppers into various segments) and *profiling data* (additional data elements that will be used to describe the shoppers in the various segments and highlight differences across segments). For the most actionable shopper segmentation, we believe driver data should be continuously available, accurate at the household level, and rooted in household behavior. For these reasons, we strongly recommend using household-level variables sourced from customer data as the primary driver variables of actionable shopper segmentation.

After the segments have been created using customer data, other data elements can then be added into the mix to profile the resulting segments. Demographics are useful to provide a sense of who the shoppers in each segment are. Modeled and/or stated share of wallet data are often *not* accurate at the individual household level, but in aggregate provide a useful sense of the sales upside with each segment. Surveys can be fielded to fill in the "why" behind the observed behaviors of each segment, and place the retailer in the context of competitive retailers.

One common concern retailers have with using their own customer data to drive the segmentation process is their need to understand the whole shopper universe within their markets – not just their current shoppers, but also their potential future shoppers. Of course, retailers do need to understand the market, and retailers should regularly field market-wide surveys to guide strategic decision-making, brand positioning, advertising strategies, etc.

However, there are two compelling reasons not to use survey responses as the driver data for shopper segmentation: One, what shoppers *say they do* and what shoppers *actually do* are often quite different. As a result, segments developed based on attitudes and preferences often show a very wide range of shopping behaviors – a much greater range than one would expect if the shoppers' attitudes were truly valid predictors of behavior. Two, survey-based segments are typically developed based on a small number of shoppers, given the high cost of fielding surveys. Many problems arise when retailers attempt to map the survey-based segments to the rest of their customer database.

Survey data is not continuously available for all shoppers, so this mapping process typically relies on the establishment of a link between survey responses and behavior, which is continuously available via customer data. This process suffers from two problems: One, keeping in mind that what shoppers *say they do* and what they *actually do* are often quite different, any link between attitudes and behavior is directional at best. Two, given the small number of households for which survey responses are available (often measuring in the low hundreds per segment), any links that can be made are typically not statistically significant and thus may not be representative of the overall shopper population. As a result, when retailers take a survey-based approach, the vast majority of shoppers (numbering in the millions) are placed into attitudinal segments on the basis of a short list of behaviors, which *may or may not be moderately linked* to the attitudes underlying the creation of the original survey-based segments (numbering in the hundreds). Given how critical shopper segmentation is to the retail organization and its trading partners, we strongly advise against basing shopper segmentation on a small number of survey responses.

What Technique to Use?

The next question retailers consider is how to bring all of these data pieces together to create distinct segments of shoppers. Which technique works best? There are two main options to segment shoppers at a higher level than basic deciling or quad-based segments: One, to extend the quadrant-based approach to more variables using a grid-based segmentation framework, or two, to use a more advanced clustering technique.

In the grid-based approach, the analyst begins by defining splits on each driver variable, then intersects these splits and names the resulting shopper segments, which are essentially the distinct combinations of the splits of each driver variable. This approach can typically accommodate only three to five variables with only two or three splits per variable; beyond this number, the resulting grid is overwhelming in its complexity.

With grid-based segmentation, results are generally "balanced" in that for most segments, each of the variables used is equally important. The main advantage of this approach is its relative simplicity, both in terms of its development (no special skill or software is required) and in terms of the ease of explaining the approach to users of the segmentation. However, there is no "magic" to this approach. The splits may or may not be the most relevant breaks, and the intersections between the different variable splits have no special meaning or importance in terms of the underlying shopping behavior.

Advanced clustering techniques, by contrast, apply sophisticated mathematics to the process to let the household-level data define the shopper segments. Clustering techniques create segments of shoppers such that differences across shoppers within a segment are *minimized* (in other words, shoppers within a segment are very similar) and differences across segments are *maximized* (in other words, each additional segment is quite different from every other segment already defined). Advanced clustering techniques can easily handle ten to fifteen different shopper inputs as driver variables.

Grid-based segments tend to be balanced. Part of the "magic" of advanced clustering techniques is that cluster-based segments tend to be multidimensional; in other words, not all variables will be equally important in terms of how each segment differs from all the others. Clustering can yield, for example, a set of price-driven shoppers who shop heavily on deal, skew to front page, and buy a lot of private label products, while also yielding a segment that is primarily distinguished by very heavy HBC/Pharmacy spending as a percent of total spend. Cluster-based segments are data-driven, and usually could not have been hypothesized by the analyst ahead of time.

The downside of using advanced clustering techniques is the complexity in the development (as these techniques require both experienced analysts and sophisticated software) and in training users on the segments. The upside is significant, however, as these techniques usually yield fewer segments that are a truer reflection of underlying differences in shopper behavior, which is why we recommend using the cluster-based approach to developing shopper segments.

A useful way to test the effectiveness of the shopper segments that have been created is to analyze a past program or merchandising change that was executed without the benefit of the shopper segmentation and evaluate whether results show meaningful differences by shopper segment. If results from a targeted marketing campaign, competitive blunting effort or a change in the deli department show "plain-vanilla" results across shopper segments, then chances are the segments won't be an effective tool to make better decisions in these areas. If results are highly differentiated by segment, you'll know you are on the right track to actionable shopper segmentation.

Implementing Actionable Shopper Segmentation

Once actionable shopper segments have been created, what's next? Most organizations would say that next steps are to train key users of the segmentation and begin using the segments to guide decision-making across functions. Yes!…but not yet. A critical step in the process too often skipped is defining a growth strategy for each shopper segment. This Segment Growth Strategy should articulate a growth target to define the overall strategy for driving growth with the segment, lay the groundwork for key tactics that will be used to execute the strategy, and identify metrics to measure progress. Segment Growth Strategies should be developed before shopper segments have been broadly communicated to category managers and CPG manufacturers. As end-users learn about the various shopper segments, the question of what to do with the segments and how to use them in decision-making is clearly addressed.

Segment Growth Strategies can be a powerful tool to overcome insight inertia, the *"Sounds great, but what's this got to do with me?"* reaction that can be a barrier to broad use of shopper segments. A sample Segment Growth Strategy for a shopper segment called "Loyal Budget Shoppers" is shown here:

Segment Marketing Strategy: Loyal Budget Shoppers

Overarching Objective	Objective	Drive sales growth by enhancing and communicating <Retailer>'s value proposition across the store on everyday products.		
	Marketing & Merchandising Tactics			
Execute the Strategy via Marketing & Merchandising Tactics	Private Label Strategy	Increase Private Label offerings (extreme value, categories)		
		Drive increased trial of Private Label across categories		
	Everyday Pricing	Implement modified EDLP program through extended TPR's and lower at-shelf prices		
		Maintain no more than a 5% gap to grocery competition on Driver products/categories		
		Communicate revised pricing (targeted & in-store)		
	Promotion	Promote relevant items via front page of circular, single-unit pricing, competitive values		
		Private Label themed wraps & Value meal themed promotions		
	Targeted Offers	Provide compelling value on Select categories via category-nominated Targeted Offer coupons		

	Key Metrics and Growth Target	Quarterly Metric	Current	Target
Track key metrics via the Segment Strategy Scorecard to gauge effectiveness	Grow spend per HH by increasing Share of Wallet	Spend per HH	$351	$369 +5%
	Increase PL category penetration	# PL Categories Purchased/HH	3	4 +33%
	Increase relevance of weekly circular	$/HH on Front-Pg Items	$71	$90 +27%
	Maintain trip frequency	Trips per Quarter/HH	16.5	16.5 n.c.
	Decrease attrition	Attrition Rate/Quarter	2.5%	2.0% -25%

The development of Segment Growth Strategies is another early test of the actionability of the segments that have been developed. If you find it difficult to articulate a growth strategy and tactics to achieve that strategy, perhaps the segments as currently defined are too broadly defined to enable highly differentiated decision-making.

In a Nutshell

The goal of actionable shopper segmentation is to put the shopper at the center of decision-making throughout the retail organization, while keeping the end result simple enough that differentiated actions can be taken in an environment in which many key levers are not targeted. As retailers move beyond the basics of shopper segmentation, they often wonder what data to use, where to begin, and which technique will best suit their needs. Spire recommends using shopper behavior, which is accurate for individual households and continuously updated and available, to drive the development of shopper segments, then overlaying as many data sources as possible, including demographics, attitudes, share of wallet and store ratings, to profile the resulting segments.

In terms of techniques, grid-based segments keep the process simple, but may not reflect true underlying differences in shoppers.

Cluster-based segments, on the other hand, are more complex to create but tend to result in fewer, multidimensional segments that are a better reflection of distinct shopper behaviors. Finally, a step too often missed is to create Segment Growth Strategies for each segment, clearly articulating the growth target, growth strategy, and tactical levers that will be used to drive incremental sales with each segment. Retailers that take these steps will be well positioned to differentiate their offerings to meet the needs of key shopper segments.

Megan Margraff is Chief Analytic Officer with Spire LLC, an industry leader in quantitative shopper intelligence, providing smart, actionable shopper analytics and consumer-centric consulting services to leading brands and retailers. For more information: www.SpireNow.com.

SECTION THREE
ENGAGEMENT

CHAPTER 6

Shopper Metrics for Optimizing Retail Performance

By Dr. Rajeev Sharma

We live in a retail world where the power has shifted to the consumer – not brands, not retailers. Consumers are well-informed, in control, and very quick in changing what and where they buy. Furthermore, due to the fragmentation of media, proliferation of products and other consumer trends, a significant portion of purchase decisions is being made in-store. Thus, understanding and influencing the in-store decision process has become critical for winning at retail. It involves mapping the in-store "path to purchase" and capturing why and how each "moment of truth" battle is won or lost.

We need measurement tools that provide direct visibility into the shopper decision process and enable shopper-centered metrics for improving every retail touch-point. Traditional research methods such as observation and survey methods have been used in an ad hoc basis to meet the need for shopper insights. Such manual methods are subjective, expensive, and under-sampled; therefore, they are not scalable. There is an urgent need for automation in the process of gathering rich shopper data to have a scalable measurement solution for optimizing retail.

Filling the Gap with Video Analytics Technology

A recent technology breakthrough by VideoMining Corporation has led to the development of an in-store measurement platform that fills the need for automatically capturing shopper behavior data. The patented video analytics technologies evolved from R&D spanning over a decade through funding from government agencies such as the Department of Defense, Department of Homeland Security, and the National Science Foundation.

The in-store measurement platform is set up without interfering with store operations. The sensors are configured to merge with the retail environment. The process involves installation of unobtrusive video sensors to cover the areas of interest in the store. Servers in the backroom of the store then process or "mine" the video streams in real time using proprietary software that converts video into rich data on shopper behavior and demographics.

The measurement process itself is totally anonymous and doesn't require any personal data from shoppers. In effect, software (instead of people) can now continuously "watch" in-store video to provide a wealth of statistical data on shopper behavior and demographics.

A key element of the technology is the ability to accurately track actual shoppers (not just carts) throughout the store. The storewide tracking enables the analysis of an entire shopping trip, providing a basis for a holistic understanding of shopping behavior including segmentation by trip types. The software also automatically derives the demographics (gender, age range, and ethnicity) of shoppers to help marketers and retailers understand segment-specific shopping behavior; for example, understanding the differences between the shopping patterns of men vs. women or how different age groups respond to specific merchandising displays.

Besides tracking the in-store "path to purchase," the platform enables in-depth analysis of the actual shelf-level interaction with products. The quantitative understanding of the key "moments of truth" provides critical insights into the shopper decisions process. The video platform captures the different merchandising and marketing elements in the store to help in documenting the "stimuli" and determining the compliance during specific tests.

The in-store video is converted into a continuous stream of behavioral data on how shoppers interact with each in-store element. The rich data is used for computing specific metrics that provide a clear understanding of shopper behavior in the categories of interest. The behavioral data is then integrated with transactional data and other available consumer data to provide a holistic understanding of the shopping process.

The central premise of this emerging shopper methodology is that "actions speak louder than words." Unlike many traditional research methods, this reveals what shoppers actually do —not what they say they do. This ability to capture unaided responses to the shopping environment results in an untainted perspective of shopping behavior. Most importantly, automation enables the objective gathering of data from a very large sample of shoppers, delivering reliable and representative data that is not available through traditional methods that rely on manual processing. It represents a revolutionary solution for addressing the urgent need for shopper understanding and optimizing retail performance.

Shopper Data and Analytics

With the measurement platform positioned in-store, shopper interaction with each element of the store can be continuously measured to extract a set of standardized shopper metrics affording visibility into key decision processes. The rich shopper data set feeds into analytical tools for quantitative and qualitative understanding of shoppers, identifying opportunities to improve brand and category performance for each market/shopper segment. This can yield marketing programs that "listen" and adapt to each shopper segment.

The data corresponding to the standard shopper metrics can be used across a diverse set of store formats, channels and categories. The metrics enables applications that range from simple filling the gap of

80 | ESSENTIALS OF SHOPPER TECHNOLOGY

"what happened in-store" to predictive modeling and optimization of all customer-facing elements of retail.

Path-to-Purchase Funnel

The Path-to-Purchase Funnel presents a set of metrics and analytics that determine the effectiveness of a category (or a display) in converting passer-bys into active shoppers and buyers. Besides rating a category's productivity, the analysis can help to identify any unusual loss in conversion ("leaks") and opportunities to improve conversion.

Traditional metrics, such as unit sales, tell us how much is sold in a given week. Loyalty card data tells us who is purchasing what products. But one of the first steps in uncovering ways to increase sales is to understand the level of traffic, engagement, and conversion. Lowering price may be effective if you know your level of engagement is adequate, but conversion to purchase is low. However, if conversion is healthy and traffic is the bottleneck, then lowering price may not be the most efficient way. A secondary display or using in-store signage may be more effective. Below are examples of the Path-to-Purchase (PTP) Funnel metrics for a major supermarket.

The Path-to-Purchase Funnel:

Store Traffic

Aisle Traffic

Penetration Rate: Category Exposure

"Category Traffic" refers to the aisle patrons in front of the section; these are patrons who have had the opportunity to interact with the products, both passers-by and actual shoppers.

Category Traffic

Conversion Rate 1: Ability to Engage

Shoppers

"Shoppers" refers to the store patrons who actively interacted with the category, including those who had visual as well as physical interaction.

Conversion Rate 2: Ability to Influence Purchase

Buyers

"Buyers" refers to the shoppers who made a purchase from the fixture

Units

$

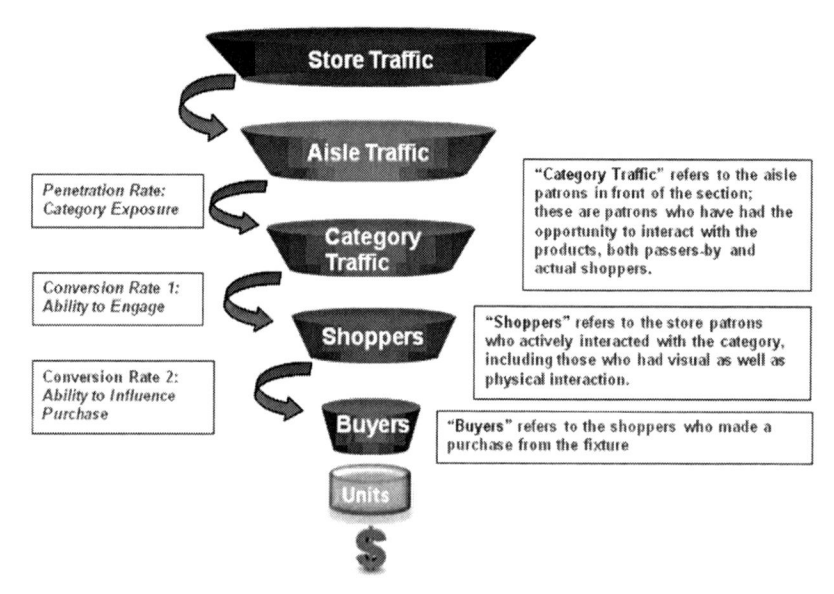

Shopper Metrics for Optimizing Retail Performance | 81

Path-To-Purchase (PTP) Metrics for Category A
- Store Traffic: 157,800
- 27%
- Aisle Traffic: 42,736
- 63%
- Category Traffic: 27,100
- 24%
- Shoppers: 6,615
- 25%
- Buyers: 1,657

Path-To-Purchase (PTP) Metrics for Category B
- Store Traffic: 157,800
- 27%
- Aisle Traffic: 42,736
- 87%
- Category Traffic: 37,465
- 10%
- Shoppers: 3,738
- 56%
- Buyers: 2,094

Category A was fairly ineffective in converting shopper engagement to purchase—only a fourth of the engaged shoppers made a purchase. Clearly, there is an opportunity to understand the purchase barriers for the engaged shoppers and find ways to overcome the barriers through appropriate stimuli – price, promotions, assortment, etc.

The path-to-purchase metrics for a second category – Category B – in the same aisle and for the same period, shows much higher exposure for Category B, but much lower engagement rate. However once engaged, shoppers converted to buyers; so the pricing and assortment seem to be optimal. The opportunity here is to find ways to engage more shoppers.

Moment-of-Truth (MOT) Analysis

This represents a set of shopper metrics and analytics for the behavior in front of a product category. The simplest metrics is the average category shopping time which, when benchmarked or indexed across other categories, provides a measure of the overall complexity of the category decision process. The shopping time is separated out from "navigation time" which, along with number of stops and other metrics, provides additional clues about the category shopability.

The distribution of shopping time (shopping time plotting against the % of category shoppers) provides key insights into the category decision process. The shape of the plot gives a quick view of whether most decisions are quick, moderate or long. Furthermore, segmentation of the shoppers into typically three to five groups based on shopping time – for example, quick, medium, long – helps in identifying the nature of decision-making in the category.

A low hanging fruit for boosting category performance is to understand the "purchase barrier" of the shopper who engaged but did not buy (the "Non-Buyer"). Once the barrier is identified, targeted approaches can be developed to improve shopper-to-buyer conversions. The approach is to analyze the behavior of the Non-Buyers quantitatively in relation to the behavior of the Buyers by deeper mining of a representative sample of category shoppers. In other words:

- How much time shoppers are browsing and then not buying?

- How many stops do they make?

- Are they going back and forth in the category?

- Are they reading the price tag?

- Are they picking up products and reading the package content?

The chart below illustrates the data from an actual supermarket category. These types of analyses help in narrowing down the principle reason(s) for not buying or the "barriers to purchase."

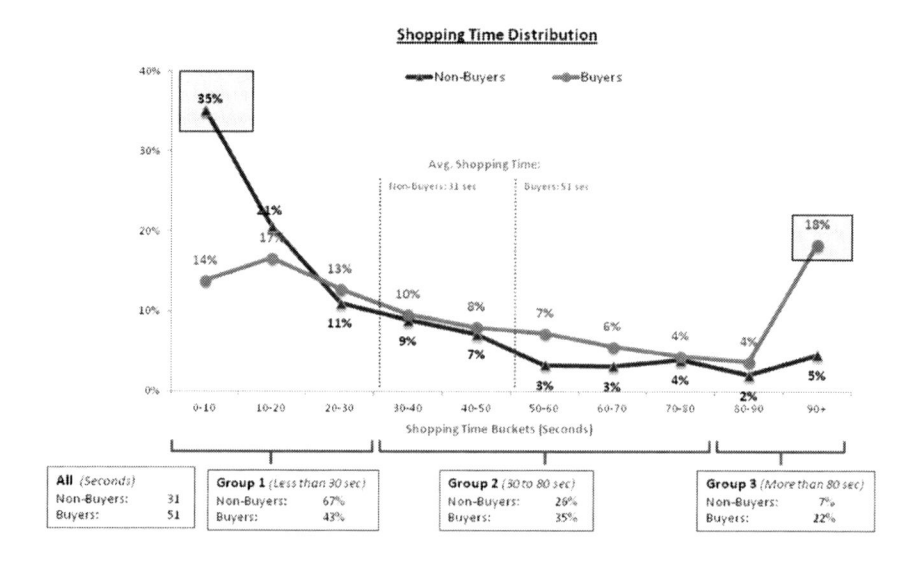

Shopping Time Distribution

Shopping Time Buckets (Seconds)	Non-Buyers	Buyers
0-10	35%	14%
10-20	21%	17%
20-30	11%	13%
30-40	9%	10%
40-50	7%	8%
50-60	3%	7%
60-70	3%	6%
70-80	4%	4%
80-90	2%	4%
90+	5%	18%

Avg. Shopping Time: Non-Buyers: 31 sec; Buyers: 51 sec

	Non-Buyers	Buyers
All (Seconds)	31	51
Group 1 (Less than 30 sec)	67%	43%
Group 2 (30 to 80 sec)	26%	35%
Group 3 (More than 80 sec)	7%	22%

Consumer Decision Trees (CDTs)

The large data set of shopper interactions with product layouts can be dissected further to build robust models of the purchase decision process including Consumer Decision Trees (CDTs). Models such as CDTs can have a significant role in guiding marketing and merchandising decisions. Traditional approaches for developing CDTs – household panel purchase data, ethnography, and shop-alongs – have clear drawbacks. They are either based on the consumer's interpretation of the purchase decision process or a human observer's point of view on decision making, providing limited understanding of the actual purchase behavior.

A patented process (US Patent Application # 12/583,080, 8/13/2009) has been developed by VideoMining to use shopper behavior data, product layout (planogram), pricing, and transaction data to develop CDTs. The key advantage of the new methodology is that it is based on actual shopper behavior as they make the decisions in-store; it is therefore more reliable. It overcomes the limitations of the conventional approaches of developing CDTs by unobtrusively capturing the actual purchase behavior of a large number of shoppers in front of the category – for example, all consumers who shop the category in a specific time period. This provides objectivity and precision not achievable through other data collection methods. It may also be enhanced by in-store intercepts for additional elements of the consumer decision process beyond overt shopping behavior.

Another key advantage of the new methodology is the ability to derive segment-specific CDTs. The large-scale behavioral data can yield a separate view of the decision hierarchy by demographics or trip missions. For example, in convenience stores, the decision hierarchy for a category such as salty snacks can be different for morning vs. lunch time vs. after work. Likewise, the decision hierarchy for frozen meals in grocery could be different by gender or age group. The competitive set or selection/de-selection process can be quite different in each case. Uncovering such differences can be very critical in developing segment-specific marketing strategies that truly match the shopper needs – whether it is new shelf-set for the category or a new product packaging for a brand.

Optimizing Secondary Displays

Secondary displays represent perhaps the most effective way to gain visibility along the path of targeted shoppers – significantly impacting performance of a brand as well as category. Despite the fact that both manufacturers and retailers have a substantial financial stake in secondary displays such as end caps due to a lack of a scalable means for measuring performance, secondary merchandising strategies are largely based on ad hoc rules with incomplete feedback on true impact or return on investment (ROI). Contrast that with marketing practices that are now standard in Internet retailing – thanks to the availability of click-through, page-view and other usage statistics.

The Video Analytics platform for capturing in-store behavior in relation to displays offers the potential to bring the Internet-type metrics to brick-and-mortar retailing. The availability of the behavioral response data enables the development of a systematic framework for evaluating and optimizing secondary displays objectively. Data from millions of shopping trips and over a large set of displays allows for an understanding of the variations for display types and different factors that impact their performance. A suite of shopper metrics and advanced analytics can be employed to enable a comprehensive and holistic understanding of shopper behavior in relation to the displays.

Video Analytics can help in measuring all aspects of how secondary displays work for particular categories and brands; for example, % volume, path-to-purchase funnel metrics, impact of product, location, types of promotion, pricing, impact for different segments, etc. Here are some sample questions about secondary displays:

- How effective is a category/brand when on display?

- What % of shoppers actually shop a category's end-cap, how long do they shop, and do they actually make a purchase from the end-cap?

- Do end-caps steal traffic from the aisle or help in attracting more shoppers into the aisle?

- How does relative placement (before/after) from the main aisle impact display performance?

- What is the optimal price point for a product when on display?

- How does a category's end-cap perform vs. other end-caps? What is the role of product type on display performance?

- Which displays are more effective for a category: end-cap, lobby, in-aisle or racetrack?

- Which categories are most effective on displays during key holidays?

- What is the optimal display configuration and design – vertical vs. horizontal, variety vs. pack-out, etc. for a category? Do displays with multiple products perform better?

- What is the role of adjacent departments on display performance?

Video Analytics can provide a deep and objective understanding of key factors that impact the performance of end-caps and other secondary displays and help develop a "play book" for optimizing secondary merchandising.

'Share of Shopper' Performance Indices

Evaluating the health of a retail business by analyzing POS sales data alone is like coaching a team by reading the scoreboard and not ever watching the full game. Any coach will tell you that it's about how you play the game that matters. You need to watch the game, player's actions and opponent's reactions to improve results.

The same is true in retail. Sure, everyone knows how many units are sold in a given week. With today's sophisticated data, we know lift over expected baseline, expected price elasticity, effects of causal stimulus such as display and feature ad space. And with the advent of shopper loyalty card data, we even know what type of shopper is buying different sets of products. Indeed, we have gotten so good at analyzing the scoreboard that we've completely forgotten about the game itself. What's happening on the field of play – inside our stores at the proverbial "moment of truth" – is all too often ignored. So, are we really watching the game?

Let's say you read sales data and found that you sold 100 units. That's great. But wouldn't you be happy to know that the 100 units sold were out of 100 shoppers who entered the store. They all made a purchase of your product! What if you sold 100 units out of 1,000 people who entered the store that day? Would it make a difference to know that of the 1,000 people who entered the store, only 200 even went down the aisle where your products were located, 25 of them did a u-turn before ever reaching your product, 25 of them breezed right by without ever stopping, 25 of them stopped for only two or three seconds, casually observing your products before pushing on, and 25 of them searched for over two minutes before walking away unsatisfied with what they saw. What if you saw the remaining 100 shoppers, who each made a purchase, deliberate between multiple brands and sizes before making the final selection? Do you think you might feel more like a coach who watched the full game? Do you think you might be able to find ways to improve your closure rate – or more importantly increase your opportunity to make a close? You bet! Welcome to the new game with shopper-centered "Share of Shopper" performance metrics.

Video Analytics enables a suite of standardized shopper-centered performance metrics for optimizing different elements of the retail game. These metrics become even more useful as they are benchmarked and indexed over all the categories and across a channel such as grocery or convenience. Sample indices include the following, with 100 being the average across all categories:

- *Exposure Index = Category Traffic / Store Traffic.* This represents the category reach or the index of exposure for a category. "Category Traffic" is the total count of all traffic in front of a category, including those who stop or those who pass by.

- *Exposure Response Index = $ or Unit Category Sales / Category Traffic.* This represents how well a category responds to exposure.

- *Engagement Response Index = $ or Unit Category Sales / Category Shoppers.* This represents how well a category responds to engagement. "Category Shopper" counts the total number of people who stop in the category even briefly.

These indices provide a good basis for comparing the key performance characteristics of each category. A benchmark provides a basis for indentifying key opportunities; for example, which categories to target for increasing exposure or which categories to target for improving engagement. The indices can be used for a component in space and promotional planning.

Tracking 'Share of Shopper' Indices

When tracked over time, the shopper indices provide a measure of effectiveness of specific initiatives and promotional efforts as well as seasonal and long-term trends. You can't manage what you can't measure. Ongoing visibility into shopper engagement and other behavior attributes allow the selection and tracking of Key Performance Indicators (KPIs) that best reflect a business goal. For example, going beyond a category, brand strength at a retail channel can be expressed as a share of the category engagement time or percentage of destination shoppers. These brand parameters can be tracked on an ongoing basis to yield a true picture of ROI for shopper marketing dollars. A host of category-related behavioral parameters can also be tracked, allowing the true performance of the category or brand portfolio to be monitored. This enables a realistic picture of the category dynamics, which can be incorporated into retail execution and into the next cycles of planning.

With the fast pace of change in both media consumption habits and shopping patterns, continual tracking of a suite of shopper metrics allows an optimized marketing execution that can rapidly respond to the changing needs of the shoppers, as well as the realities of the retail execution. The shopper data suite from in-store video provides an attractive platform to build and activate brands at retail, communicate with consumers in the store, and transform the retail experience.

Bringing Shopper Focus to Category Management

The new suite of shopper metrics allows the category management function to systematically include the shopper in the center of all its activities.

- *Scorecarding* – To go beyond unit and $ sales, market share, and category segment performance to include category exposure, engagement and conversion rates along the Path to Purchase.

- *Efficiency Analysis* – To go beyond sales per foot, pack out, and days of supply to include efficiency measures such as Exposure Response, Engagement Response, and $ sales/minute.

- *Pricing and Promotional Analysis* – To go beyond Price Elasticity, Transferable Demand, Causal Analysis, Base/Incremental Analysis, and $/Transaction to change in exposure, engagement and cross-shopping.

The new shopper metrics provide an opportunity to frame and address burning category questions, such as:

- Do we have the optimal category schematic, shelf flow and aisle agency?

- What is the impact of a new shelf re-set or product flow on shopper behavior?

- What is the optimal end-cap configuration and why?

- How do shoppers actually shop my category and can we better serve their needs?

- How do shoppers engage with in-store marketing activities such as POS or coupon machines?

- How can we convert more category traffic into buyers?

- Does our category receive its fair share of store traffic?

- How does our category's traffic and conversion compare to other categories?

Real World Learning Labs

With visibility into how shoppers respond to specific marketing elements, testing can be elevated to a whole new level. Access to precise measurement of behavioral responses in both pre/post and test/control modes can help enhance any in-market testing by getting to the "why" and "how" behind sales numbers. The extra visibility helps with diagnostics and support robust, fact-based decisions, improving the ROI for any new initiative. The shopper metrics, for example, can pinpoint stages of the "Path-to-Purchase Funnel" where the conversion has room for improvement.

Thus, the shopper marketing mix and parameters can be refined to maximize the chances of success. For example, if an in-store POP display is one of the vehicles chosen, the relative placement of the display with merchandize can greatly impact shopper engagement, especially for new products. A recent project demonstrated this effectively when changing the location of a POP display by a few feet at a mass retailer led to 300% improvement in engagement levels. The engagement measures are also crucial in fine-tuning brand attributes highlighted in the packaging. Furthermore, testing can provide insights into the impact on category dynamics, which is very helpful for supporting retail execution.

Innovation is the key to winning at retail. And the key to winning at innovation is to first understand the barriers to purchase, then to test solutions that break through those barriers by closely matching the shopper's needs, and finally to track results over time. Manufacturers who work collaboratively with their retail partners and use in-store measurement and analytics can make better recommendations and take action rapidly and with greater confidence!

In a Nutshell

There is an increasingly urgent need to understand the ROI of shopper marketing spending and to find the best mix of stimuli to influence the purchase. Video analytics technology enables clear visibility into the Path-to-Purchase and all key Moments-of-Truth in the store. By using the in-store shopper behavior data and analytics, marketers can have a principled and scalable way to understand opportunities, test innovation and track results in real-time.

Recent advances in technology have made it possible to map the "path the purchase" accurately and analyze each "moment of truth" to precisely understand the factors impacting the purchase decisions or barriers to purchase. Real-time data streams from every second of every shopper (much like the web analytics) are now revolutionizing the way marketers make informed decisions about every aspect of trade promotions, category management and shopper marketing.

Access to real-time shopper metrics can help:

- Uncover opportunities for enhancing retail performance by fact-based analysis of the shopping process.

- Test innovative merchandising/marketing concepts rapidly by quantitative measurements of shopping behavior by target segments.

- Monitor impact of promotions by tracking category penetration, conversion and other KPIs in real-time.

Dr. Rajeev Sharma is the Founder and CEO of VideoMining, a leading provider of shopper marketing insights using technology-based in-store measurement and analytics. Dr. Sharma is recognized as a pioneer in human behavior analysis and shopper marketing research. Prior to founding the company, Dr. Sharma was a tenured faculty member at The Pennsylvania State University and a research faculty member at The University of Illinois at Urbana-Champaign. More information: www.videomining.com.

CHAPTER 7

The Evolution and Application of Virtual Shopping: Past, Present and Future

By Andrew Reid, Matt Kleinschmit and Richard Rizzo

The practice of employing life-like, simulated testing environments for research purposes is nearly 30 years old. There appears to be no slowdown in evolutions to both technology and methodologies employed in virtual shopping exercises. The past decade alone has seen significant advancements in visualization technology, user interfaces and, perhaps most importantly, the applications for which virtual technology are employed. From fully immersive virtual shopping environments to recent trends in gamification, researchers are rapidly attempting to make their surveys as visual as possible to enable a more realistic research environment that promises to both engage consumers and provide better quality data.

Companies have increasingly turned to virtual testing solutions to assist in identifying how consumers shop specific categories. They aim to optimize marketing, product and package variables to position products better in the decision hierarchy. The end result of the research provides a framework for engaging consumers with products, planograms and communications strategies that best meet their needs and maximize their Path to Purchase experience. It also can help to engage retailers to make proactive decisions on merchandising and assortment strategies that can lead to improved sales in critical product categories.

Despite the increasingly vast array of virtual reality-based research that is occurring, there is surprisingly little existing research on the effectiveness of various virtual testing techniques. Clients routinely ask about the impact of varying virtual testing technology and research methods. For many research suppliers who offer virtual testing services, the answers to these questions are often predictably biased to support the type of virtual testing services they offer.

This chapter traces the evolution of virtual testing and explores both current technology usage and research applications across a wide variety of industry verticals. In addition, we examine current industry best practices for how virtual technology tools can be leveraged to address market research questions, and identify those methods which generate actionable insights most efficiently. This includes the impact of 3D vs. 2D virtual technology, monadic vs. sequential monadic research designs, smaller vs. larger categories and traditional vs. emerging virtual eye tracking techniques. Finally, we explore emerging mobile technologies such as augmented reality and discuss ways in which this technology could be integrated with virtual reality to create powerful and targeted research applications.

There is no doubt that virtual technologies will become even more common in the years to come. Advances are making the experience even more immersive with applications to a wider range of situations. But as technology evolves, so too must our research skills and the role of researcher as an arbitrator of this technology in knowing how best to apply it when addressing key business issues. As we will see, the objective of the research should determine the scale of virtual complexity used, and good research doesn't necessarily need all the bells-and-whistles. In fact, sometimes more basic versions of the same virtual environment can deliver comparable results with less time and financial outlay. In the end, virtual testing is another tool in the insight-gathering toolkit, and its benefits and limitations must be fully understood for effective real-world application. This will be particularly true in the coming years as both visual and virtual engagement with consumers continue to expand in both research applications and industry use.

The Evolution of Virtual Testing

From its origins in the 1980s, virtual testing techniques have captured the imagination of marketers around the world. The ability to leverage simulated virtual environments to test new products, marketing communications and retail experiences has been profound, fueling a desire for greater flexibility and validity in practical research applications. For many marketing researchers today, virtual testing is an engrained activity. In fact, many traditional methodologies are now using facets of virtual technology in subtle ways – from "visual questions" and more realistic three-dimensional product images to fully immersive virtual shopping environments.

Early forays into simulated virtual environments employed technology available at the time. Professor Raymond Burke, E.W. Kelley professor of business administration and renowned pioneer in the field of virtual testing, began working with simulated environments in the late 1980s at the University of Pennsylvania's Wharton School. In those days, virtual testing employed large-scale video to mimic walking through an actual store; however, the collection of data was still done via interview or paper-based surveys. As computer technology progressed, so did the level of complexity with virtually simulated environments. The interactivity of the computer – and eventually the internet – allowed for the collection of significantly more data without interrupting the participant to ask questions.

Many companies have been experimenting with computer simulation since the 1990s. These include industry icons like Procter & Gamble, PepsiCo, Intel and Kraft – organizations that felt that 3D computer graphics were and are practical test-marketing tactics that can change the way innovation and strategy are approached and executed. More recently, both manufacturers and retailers alike have experimented with developing specialized stand-alone facilities with fully immersive 3D virtual testing equipment to continually test new retail environments, in-store marketing materials and category arrangements.

A trip down the evolutionary path of virtual shopping reveals three major milestones that have brought us to the present state. The following milestones also represent significant benefits both to the participant and the user of the research:

From Flat Images to 3D Modeling. Early simulated environments were created with actual video footage and such video required significant computer power to display and control. As technology progressed, environments could be rendered graphically; however, the *reality* of the environment could be questionable. More recently, virtual reality has really taken flight, especially with the advent of computer gaming and real time video rendering. Products, objects, lighting, etc. look almost like real-life with incredible depth and texture. Modeling in 3D provides true interactivity, allowing participants to demonstrate behavior just like in a real space. This also provides for the collection of a vast array of behavioral data (for example, time spent shopping, items looked-at, etc.) not just *what* was purchased.

From Super Computer to Notebook. Historically, virtual simulation was so computationally intensive that it could only be handled by advanced high-powered computers. Such computing power was a limiting factor only available to a few manufacturers with deep pockets or rogues with a desire for innovation. Similarly, such technology required large spaces to house the testing facility and the processing computers. Today, virtual environments are available locally – right on a desktop or notebook computer. Internet technology (HTML), Flash, Java and other advanced program languages make the creation and deployment of virtual environments practically ubiquitous. The obvious benefits are that the environments are not only easier to create, but they can be deployed anywhere at the participant's convenience (not the researcher's).

From Central Location Testing (CLT) to Online. Some would argue that face-to-face research is superior; however, still others would argue that larger sample sizes derive the greatest confidence and quantitative rigor. Well, virtual technology allows for both. When virtual environments needed vast computing power and large spaces to house the expensive equipment, testing was limited to central location. While CLT provided an opportunity to *hear* the consumer, today's online methods provide significant opportunity to reach large, nationally representative samples of consumers (including key segments of interest) and still hear what they are saying plus *see* what they're doing. Furthermore, online virtual shopping methods provide anonymity, convenience and turn-around that are simply not available via CLT.

Typical Virtual Shopping Functionality

Current virtual shopping technology works much like online shopping interfaces except that respondents are presented product options on a simulated, interactive shelf that mimics the same section in a real-life store. Respondents simply scan the shelf by navigating with their mouse to see the products. Products can be picked up, turned over, placed in the shopping basket or returned to the shelf. Even products initially selected for purchase can be returned to shelf, with the final stage involving the respondent purchasing the items at checkout. A key benefit of virtual shopping is that all of the event-based behavioral data is collected during the exercise. In some cases, subsequent survey questions can be linked to shopping behavior to further exploration shopper attitudes and motivations. Following is an example of VIsion Critical's virtual shopping interface:

Vision Critical Virtual Shopping User Interface

Validation

As with any new technology, there are barriers to acceptance: fear, confusion and complacency – to name a few. However, the vanguards of the movement have transformed the use of virtual environments from a fad for curiosity into a tool for proactive, ongoing brand and category management. In fact, real world validation is well documented and best practices are being developed and adopted. Some examples of recent validation include: projection of volumes from virtual shopping data that are in line with syndicated data after the test, planogram optimization and real world sales increases, package testing leading to increased consumer acceptance, etc.

The reality of virtual is that it no longer remains the stuff of science fiction. It is a real and present tool for marketers and researchers. Risk takers and innovators took and expanded on the fact that behavioral data (what people do) is much better than articulated data (what people say) to create a platform for insights generation that leaves old-school paper surveys in the trash can. As technology and comfort with virtual testing expands, the applications will only increase. The only limits will be the creativity and ingenuity of marketers and their counterparts in consumer and shopper insights.

Today's Virtual Testing Applications and Technology

In the 2007 report by the Grocery Manufacturers Association/ Deloitte Consulting called "Making Shopper Marketing Work," virtual shopping was praised for "allowing for rapid and realistic scenario testing of merchandising, product and promotion designs and layouts with reduced need for field testing." Indeed, these applications are now widely used within the Fast Moving Consumer Goods (FMCG) industry, which has rapidly integrated virtual shopping into its extensive shopper insights and category management toolkits. Virtual shopping data is routinely gathered by FMCG manufacturers and shared with retailers to help support new product placements and aisle flow changes. In response, many retailers are not only embracing and leveraging these data in their decision making, but also requesting additional information and strategies from forward-thinking manufacturers.

Current and Emerging Virtual Shopping Applications

Probably the single-biggest driver of growth in the virtual shopping arena over the past 15 years has been from retailers and manufacturers working together to improve the performance of individual FMCG categories. "Category Captains" (those manufacturers who represent the largest share of category sales) are routinely tasked by retailers with developing strategies that will invigorate growth. As a result, many manufacturers have turned to virtual shopping methods to help test a wide variety of category management strategies and tactics. But the uses and applications of virtual shopping are expanding as creative marketers and researchers look to leverage the considerable benefits of immersive virtual testing stimuli into a wide array of product innovation, portfolio optimization and shopper exploration initiatives. Below are a few examples of both current and emerging virtual shopping applications:

Category Reinvention. While many FMCG manufacturers routinely test the effectiveness of alternative product and brand blockings in key grocery categories such as cookies, frozen dinners, etc., they are also increasingly expanding this critical application of virtual testing into broader category reinvention initiatives that seek to optimize entire category flow and sub-category adjacencies. In a study recently completed by Vision Critical and reported on in Shopper Marketing

Magazine, a leading FMCG manufacturer was interested in re-creating the ever confusing condiment aisle from scratch, incorporating shopping behaviors and preferences into new planograms for both grocery and mass channels. This study leveraged a new, interactive virtual category creation exercise called "Build Your Own Aisle," that allowed respondents to create their most-preferred aisle flow and sub-category adjacencies.

The results of this online quantitative exercise were used to create two category configurations which were subsequently tested via online virtual shopping along with other manufacturer- and retailer-driven aisle scenarios. Interestingly, the planograms that were derived from shopper-preferred adjacencies produced the greatest incremental reach, revenue and volume for both the client's brand and the category overall. This prompted a key national retailer to begin in-store testing of these new shopper-preferred category arrangements to validate the results prior to broader implementation.

Portfolio and Pricing Optimization Research. Global macro-economic trends over the past several years (including rising commodity prices) have prompted a rise in the use of virtual testing methods to test new portfolio, pricing and product size strategies. In some cases, virtual shopping is being integrated with Discrete Choice Modeling (DCM). The associated simulation tools generated by these methods are used to help predict how consumers will respond to changes in package size or price (or both), and how these portfolio changes affect reach, share, volume, profitability and product migration. FMCG manufacturers leading these efforts believe that the investment of virtual shopping-based category decision-making allows them to help guide retailers on proposed category changes. It also positions them as valued partners in ensuring incremental category growth, while also ensuring that any changes recommended affect their own brands in a positive fashion.

Packaging and Design Research. As packaging continues to play an important role in standing out on shelf in an increasingly cluttered retail environment, many manufacturers are embracing virtual testing to move beyond a "beauty contest" of design to actually quantify how new packaging, graphics, structure or the inclusion of other elements such as secondary and shelf/retail-ready packaging affect at-shelf shopper behavior. This is particularly true as more manufacturers are shifting to more cost effective packaging (such as pouches) to help offset rising raw material costs. When armed with virtual shopping data that shows the impact of changes on actual purchase behavior, companies are less likely to miss the mark when it comes to real-world execution.

Point-of-Sale (POS) and In-Store Marketing Research. Both manufacturers and retailers have adopted virtual shopping methods to understand what impact various signage, secondary displays, end caps, and other in-store Point-of-Sale (POS) materials have on sales. In this context, virtual shopping technologies allow multiple creative executions to be tested virtually prior to deployment, providing an in-depth understanding of how various creative and communications tactics impact shopping and purchase behavior.

Disruptive Retailing Strategies. As manufacturers look to non-traditional distribution and merchandising strategies to help increase incrementality (such as secondary merchandising locations, vending and other scenarios), virtual testing methodologies are being employed to predict the performance of secondary distribution and merchandising scenarios efficiently and also to assist in "sell-in" presentations with retail partners. Armed with virtual testing data that quantifies the impact of these strategies, manufacturers are increasingly targeting specific outlets such as mall operators, retailers, Quick Service Restaurants, etc.). They are providing them with fact-based evidence as to why distribution through their channel or secondary merchandising locations makes sense.

Virtual Shopping-Based In-Home Usage Test (iHUT). Manufacturers have long embraced iHUTs to assess product performance and repeat purchase. They are now also beginning to book-end these studies with virtual shopping exercises so that the impact of the product's performance on reach, revenue and volume can be quantified. This new type of virtual shopping-based iHUT consists of an upfront virtual shopping exercise followed by an in-home product placement and evaluation based off of the products actually purchased in the virtual shopping exercise. Then, a virtual shopping exercise among the same consumers later on is conducted to quantify the impact of actual product usage on subsequent shopping and purchase behaviors. This unique longitudinal research plan provides a multitude of data for fact-based product and category decision-making for both manufacturers and retailers alike.

New Product Category Placement. As new product innovation continues to grow, FMCG manufacturers are also using innovative virtual shopping applications to not only understand how a new product will perform on-shelf and where its source of volume may come from, but to also understand where in-store a new product should be placed. This is particularly true for disruptive product innovations that have the potential to be shelved across a wide variety of categories. In those cases, manufacturers are employing hybrid virtual testing methods to both assess benefit alignment with other similar products and to quantify performance when placed within a variety of categories.

In-Store Eye-Tracking. As in-store environments are becoming increasingly cluttered, both manufacturers and retailers are seeking more information on what catches shopper attention – from end cap displays and shelf violators to package designs and POS. While eye-tracking techniques have traditionally been used to answer many of these questions, new alternative eye-tracking methods have emerged in which research participants are shown virtual stimuli and asked to click on areas that pique their interest. While these methods are more cognitive in nature, they are often accompanied by interactive, online reporting tools that allow for diagnostic refinement of stimulus to maximize their effectiveness.

Broader Industry Adoption of Virtual Testing

Just as the applications for virtual testing have expanded within the FMCG vertical, other industries are also embracing the promise of simulated environments and interactive tools within the consumer research they conduct. Here are a few noteworthy examples:

Technology and Telecom. The technology and telecom industries routinely use virtual testing for understanding in-store shopping behaviors, optimizing pricing, and identifying which Point-of-Sale (POS) information is most impactful on purchase. This research often includes multiple cells of respondents with each exposed to different product information and pricing scenarios, with the resulting comparative analysis revealing the relative impact of varying levels of product information on purchase intent.

Financial Services. Banks and other financial institutions have turned to virtual testing approaches for the same reasons as retailers: to understand guest experience and identify areas within new branch designs and in-store marketing materials that may influence consumer behavior. Over time, this has evolved into the development of new approaches that serve to simulate or mimic other consumer touchpoints with their financial institution.

Automotive and Aviation. Automotive and aviation manufacturers have also experimented with a wide variety of virtual testing methods for optimizing the design and interaction aspects of their products. This includes immersive virtual depictions of exterior and interior design elements in which navigation/ viewing behavior is tracked and the virtual experience is followed up by evaluative questions surrounding consumer preferences.

Quick Service Restaurants (QSR). Quick Service Restaurant operators have also adopted virtual testing methods for interior and drive-thru menu board optimization, as well as in-store signage testing. In these applications, virtual ordering experiences are mined to identify which menu items and signage are most impactful in purchase decisions, allowing QSR operators to optimize menu real estate to maximize sales of high-profit items by day-part and shopper types.

Others. A wide variety of other industries are also experimenting with virtual testing methods, including casinos for new game testing and shopper marketing agencies that are using virtual testing to validate in-store and creative executions of their work.

The Future: Enhanced Uses of Virtual Reality, Augmented Reality, and Beyond

As we have seen in virtual shopping's evolution over the past ten years, it is highly likely that the technology and its applications will continue to expand in the years to come. Already we are seeing advancements in the ability to deliver even more realistic virtual simulations via online studies, as well as the integration of virtual stimulus with other complementary technologies such as biometrics, neuroscience and facial recognition.

One area that holds significant promise in virtual testing is the integration of advanced Smart Phone Augmented Reality applications with immersive virtual stimulus. While virtual reality is the exercise of creating realistic environments in a virtual setting, Augmented Reality is defined as a live, direct or indirect, view of a physical, real-world environment whose elements are *augmented* by computer-generated sensory input such as sound, video, graphics or GPS data. This could signal a turning point in research because we can now expose shoppers and respondents to "what if" scenarios while they're in-situation, effectively merging real-life, augmented and virtual environments together to achieve specific research objectives. By changing or "augmenting" a shopper's reality, researchers can expose them to situations while they are in full shopper mode, in a real store, and see what effect this has on their behavior.

In research applications such as package testing, respondents would use their smartphone to scan over a specific product in-store. That would trigger the application to start a short mobile-based survey that tests alternative packaging designs while the shopper is still at shelf and in-store, gathering data on their reactions to the new packaging. Findings are subsequently reported in real time to incent and encourage future respondent participation. Below is an example of how could work:

Vision Critical's Augmented Reality Application for Package Testing

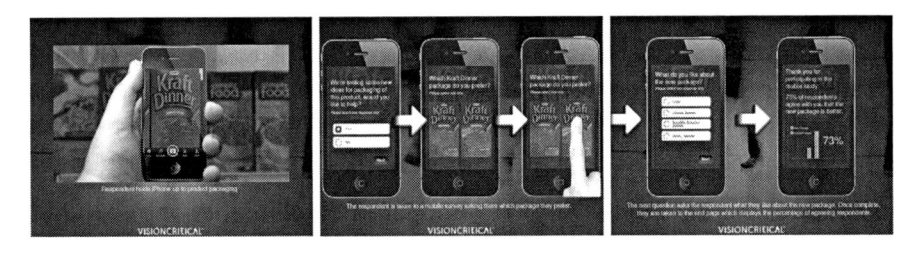

Similar applications of this technology could include testing of alternative in-store signage and POS materials, much like the packaging example above, followed by shopping exercises to help understand the impact of such scenarios on purchase, volume, trial and switching behavior.

Integration of augmented reality research with GPS-enabled location-based services will also gain prominence as a research tool in the near future, allowing both manufacturers and retailers to test new products, category arrangements and in-store marketing materials, and extending into adjacent industries such as hospitality, sporting and entertainment venues where consumer touchpoints are typically not maximized to their fullest potential.

Longer term, augmented reality and image recognition technologies (such as Google Goggles) may help marketers to capture and quantify brand touch points throughout consumers' daily lives. This holds promise in reinventing brand health tracking initiatives, replacing brand recall with actual tracking of exposure (including location) and truly understanding both the incremental and cumulative value of brand interactions on favorability, health and purchase.

Similarly, this type of application could also replace consumption/ purchase diaries and product scanning methods, allowing consumers to simply photograph products they purchase or consume in a day, or capture an image of their pantry, versus the manual nature of the traditional diary and inventory data collection methods.

Summary and Conclusions

Virtual testing technologies are in a perpetual state of motion with new developments rapidly creating exciting possibilities for researchers around the world. And as these possibilities expand into new research applications across industries, researchers will be increasingly tasked with ensuring that the appropriate technology is deployed to answer the business issues at hand. Mobile-based augmented reality applications, facial recognition, GPS enabled location-based services and advanced 3D gamification techniques all hold promise as viable techniques that could provide greater flexibility in research design and applications.

But as the market research industry applies these technologies in the coming years, careful consideration must be taken to continually leverage the same techniques that we would employ to help our clients to verify whether a new method is viable and accurate. Without rigorous testing to understand the impact of varying technology and methodological designs on data quality and insights, the industry risks losing credibility in effectively leveraging technology and could cede this position to outside industries.

Our hope is that the market research industry as a whole can benefit from collaborative exploration on how best to deploy emerging virtual reality technologies to address key business questions. Clients will certainly expect an increasingly visual and virtual research experience, and consumers will indeed require it in exchange for their attention, opinions and engagement.

References:

Breen, Peter (2009). *"Shaping Retail: The Use of Virtual Store Simulations in Marketing Research and Beyond."*

Holston, Rob, et al (2008). *Making Shopper Marketing Work. Deloitte Consulting/Grocery Manufacturers Association.*

Urbanski, Al (2011). "Clorox Makes Consumers Category Captains," *Shopper Marketing Magazine*

The Authors:

Andrew Reid is the founder and Chief Executive Officer of Vision Critical, a provider of online market research platforms and services that facilitate two-way communications with customers, employees and citizens for a diverse global client base. Matt Kleinschmit and Rich Rizzo are the respective Senior Vice President and Vice President within the company's Integrated Consumer, Shopper and Retail Insights division. For more information: www.visioncritical.com.

SECTION FOUR
ANALYTICS

CHAPTER 8

The Evolution of Trade Promotion Management

By Lorne Schwartz

Consumer Packaged Goods (CPG) manufacturers rely heavily on the use of trade funds to proactively shape demand. The average shopper may not be aware of it, but virtually every product placement, price reduction and end-cap display has been funded by the manufacturer. The objective of such programs is to influence both retailers and consumers. For example, trade promotion remains a critical tactic for getting more shelf space and helping retailers increase store traffic. It helps create product and brand awareness among consumers and can stimulate trial purchase and switching behavior.

Trade Promotion Management (TPM) is the management of the entire cycle of trade promotions and related spending between CPG manufacturers and their retail customers from annual planning, creation, implementation and post-event analysis to settlement. For the average CPG manufacturer, trade spending ranks second on the company's P&L. And despite the recent recession, trade spending has not decreased at most companies. Although the volume of spending on trade promotions is considerable, research shows that CPG manufacturers still struggle to find the right technology solution for this critical business activity. Legacy financial management systems are well entrenched across the industry, but those applications are not built to handle the more specialized processes related to trade promotions. As a result, over half of all CPG companies today rely on homegrown spreadsheets and custom-built tools.

Unfortunately, today's homegrown tools are struggling to meet management reporting requirements. As competition for retailer and consumer attention increased over the years, trade promotion spending increased tremendously. Studies consistently show that it often accounts for 10 to 30 percent of gross revenue. Not surprising, executive management, the Board of Directors, lending institutions and even outside investors have taken a great interest in this line item. Accurate and timely information is now in great demand as many of the homegrown approaches are failing to keep up.

Ongoing Power Struggle

The current economic environment has given retailers increased control over their relationships with suppliers. Historically, the balance of power between the manufacturer and the retailer has vacillated, but in the present economic environment, the pendulum has swung far in the direction of benefiting the retailer. Many CPG manufacturers have responded to the increased pressure by cutting prices. This may increase demand in the short-term, but it is not a sustainable, long-term strategy.

Small to midsize CPG manufacturers are now feeling the brunt of these recent demands even more acutely. They often have less influence with the retailers they work with because they are not the category captains. Also, historically they simply cannot afford access to data that would help them make more informed decisions. This leaves them in the difficult position when dealing with retailers and other key stakeholders.

An ongoing nightmare that CPG manufacturers have about their retailers is unexpected deductions. Some manufacturers will choose to not challenge a deduction if it is under an established dollar amount.

In their minds, it takes fewer resources to clear it without further examination; they simply cannot afford to spend time and effort fighting it. Since they are not category captains, there's even a fear of damaging the relationship if they do push back. The power lies with the retailer, with the manufacturer leaving money on the table.

Furthermore, with consumers starting to shift to purchasing private label brands and retailers implementing SKU rationalization to essentially cut out national brands, strained manufacturers need to know what kind of return they are getting on their sizable – and increasingly precious – trade spending investments. Without good information, a manufacturer will struggle to make the right choices.

State of Homegrown Spreadsheets

Without a single, centralized system of record for promotions, CPG manufacturers lack visibility throughout their organizations or between departments, often resulting in frequent surprises and fire drills. Many companies simply conclude that with spreadsheets, measuring promotion effectiveness is almost impossible. Critical downstream data sources come in different formats and are stored in different locations. Sales data is locked away in the ERP system, while external data might live in a portal or sit on a server in raw flat file form. Companies managing their promotions through spreadsheets have no quick or easy way to import these data sources to match against the promotion plans. As a result, valuable information often goes unused.

Even monitoring trade spending is a challenge. When promotion plans and sales forecasts reside on local spreadsheets, there is no easy way to link the two together. For example, if an individual sales manager increases a forecast number or changes the terms of a promotion, the software has no mechanism for automatically calculating and then communicating the spend liability back to finance.

After promotions run their course, accurately reconciling deductions and bill-backs become expensive hassles. Manufacturers that store their promotion plans in spreadsheet format create a real challenge when trying to tie a deduction back to the right promotion. Over time, these companies could be losing millions of dollars.

By using spreadsheets, companies often don't even know if a promotion is profitable. Promotion profitability requires good data to calculate. Cost of Goods Sold (COGS) and downstream data both need to be assembled to determine all the costs associated with product sales as well as to track actual prices at the point of sale. Since this information is dynamic and often inconsistent, it can be very difficult to assemble it manually in spreadsheets on a regular basis. This is one of the leading reasons that promotion profitability is the most important – but least used – metric by companies today.

Spreadsheets are wonderful for dissecting a known data set, but fall short of being a true business intelligence tool. While pivot tables can assist with multiple dimensions of data, spreadsheets don't handle data integration, transformation, consolidation or hierarchical relationships well. Unfortunately, these are all critical requirements for effectively managing trade promotions, making spreadsheets ill-suited to perform the job.

Determining True Costs

When trying to decide if the time has come to invest in a more sophisticated TPM application, most companies begin with a cost/benefit analysis. Unfortunately, many of the companies underestimate the cost of their current approach. A good analysis must explore the following:

- **Labor Costs.** Perhaps the most frequently ignored element of spreadsheet cost is manual effort. Consider how much time is spent building, maintaining and securing spreadsheets for each sales employee for a month and add it all up. The resulting sum is certainly in the order of weeks, not hours or even days.

- **Opportunity Costs.** Closely tied to labor cost is opportunity cost. What are the activities your sales employees could have been engaged in had they not been fixing or managing a spreadsheet or a report? In CPG, sales people need to report back to their retailers and their managers regularly – often in dozens of different formats. It is not uncommon for this to take three to five days per month simply managing different reports from a single source of data. Isn't that time that could have been better spent analyzing the business, doing store audits or collaborating with buyers?

- **Security and Risk Costs.** While risk can be very difficult to quantify, the "CFO test" can be a good substitute here. For instance, what would your Chief Financial Officer think of the idea that he or she has no way of knowing about a huge spike in trade liability for the coming month? Or the idea that an entire sales plan could be left in a hotel on a lost laptop? What about a post-audit, where the back-up documentation is all stored on the laptop of someone who left the company a year ago? While you may not be able to pin a cost to risk, the CFO's typical reaction to these risks says it all.

- **Cost of Human Error.** Too often, a single missed keystroke in a formula or an accidentally deleted cell can mean that entire workbooks or spreadsheets are wrong. How confident are you that all the data in the sales team's spreadsheets is accurate and reliable? This is information they are using to make key decisions about promotion strategies, communicate spending liability and measure promotion effectiveness. You wouldn't manage your Cost of Goods Sold (COGS) in a spreadsheet, so why would you manage your second biggest expense that way?

- **Cost of Scalability.** Scalability remains one of the most common triggers for a CPG company to move away from spreadsheets and onto true centralized enterprise software. Most companies start out small, and they can manage their business with basic desktop tools. Then centralized ERP comes on board, and soon they are growing by double digits. The effort behind managing and consolidating spreadsheets becomes exponentially more difficult until finally processes start to break down. In many cases, a lack of an enterprise trade management system can singlehandedly prevent a CPG firm from growing into larger national accounts. That can represent a significant opportunity cost in addition to growing labor costs.

Shifting the Paradigm

CPG companies need better information to gain back some of their leverage. They need to be able to show how certain promotional events benefit the retailer – such as increasing store traffic. They need to be able to determine sales and profitability accurately. They need to understand the ROI of every retailer and every promotion. They need to be able to measure, analyze and predict quickly and precisely.

Fortunately, times have changed and technology has brought companies a practical and affordable solution. Cloud computing has come to the world of trade promotions. CPG companies of all sizes now have access to Software-as-a-Service (SaaS) applications built specifically for the trade promotions process. This type of application is typically hosted by the service provider and requires no physical software implementation. There is also no investment in hardware or middleware and little-to-any IT support is required. Maintenance, upgrades and client service is often included in one simple monthly fee.

What's Next on the Evolution Path?

Over the past several years, price and promotion optimization have become the hot topics among consumer products manufacturers looking to get greater value. Especially in a tight economy, few can afford to gamble and lose on their price and promotional tactics. But while the concept has become mainstream, the definition, requirements and scope remain murky. A solid foundation of transactional systems, data collection and sound investments can move a CPG company in the right direction without draining IT budgets in the process. *This is called Trade Promotion Optimization (TPO).*

TPO applies advanced predictive models to multiple downstream data sources to make the promotion planning process more scientific, predictive and accurate. TPO can forecast base volume by customer or event and forecast total volume by week, by pounds, by gallon, etc. It can also optimize proposed events by field sales and evaluate promotions based on "true net lift." In short, it helps make a manufacturer's trade funds spending more effective while increasing the return on investment for a TPM solution.

Let's apply this to a real-life example. A sales team might ask the question, "If we promote this item at this retailer on this date, with a price reduction of $X, a feature in their circular, and a display, what will be the lift?"

With optimization, a team can instead ask, "Assuming that we have $50k to spend with this retailer and that we are going to promote this item in this time period, which type of promotion, or combination of promotions, will provide the most profit for both parties?"

In the first example, the promotion types are the inputs, whereas in optimization, the process is essentially reversed; the promotion types become the outputs.

The key with TPO is starting with good data. Having a solid foundation of historical trade promotions is crucial to implementing a useful TPO solution. Incomplete data will not permit accurate predictive analysis. Furthermore, data must be synchronized and harmonized across channels. The more granular the data, the more granular the optimization. Before attempting TPO, it is imperative to have a successful TPM solution in place.

TPM: Out of Excuses!
As Trade Promotion Management has grown in importance, so has executives' frustration with their inability to track, manage and predict the effectiveness of trade campaigns. Spreadsheets and homegrown tools worked for a while, but they can't keep up with the evolution of

the industry. Technology and best practices have come to the rescue. CPG companies of all sizes can afford state-of-the-art SaaS trade promotion solutions. These solutions can literally be up and running in weeks, not months. With financial reporting and compliance pressures at an all-time high, the timing couldn't be better. Now there are no excuses to act! The benefits will certainly not only be limited to more accurate and timely information, better communication and more confidence in the decision. A good TPM system will save money in several ways: elimination of unprofitable promotions, identification and replication of profitable promotions, improved forecast accuracy, superior deduction management and improved trade liabilities.

Seven FAQs about Trade Promotion Management

What is Trade Promotion Management (TPM)?
TPM is the management of the entire lifecycle of trade promotions and related spending between Consumer Packaged Goods (CPG) manufacturers and their customers. The lifecycle spans from annual planning to post-event analysis to settlement.

Why is TPM an important software category for CPG?
Trade promotions are expensive and difficult to track and measure. For most CPG manufacturers, trade spending represents the #2 line item on their P&L, often 10-30% of gross sales. This cost impacts every part of the organization from sales, marketing, manufacturing, finance, operations and supply chain. Using a manual process, manufacturers have difficulty measuring spend performance, determining promotion profitability, and reconciling deductions from retailers and distributors.

Why do I need software to do these things?
Many companies don't track their trade dollars, or they use inadequate tools. CPG companies still use desktop spreadsheets as the primary source to track promotion data. Without a single system of record for

promotions, companies lack visibility up and down the organization or between departments, often resulting in frequent surprises and fire drills.

Why Should I Use a TPM Solution?

A TPM Solution creates a single, centralized system of record for trade promotions and related spending. This gives common access to all departments, supplying them with consistent, accurate and current information to help make better business decisions. Sales managers can spend their time selling product, not compiling spreadsheets and reports. Corporate finance can immediately access the most up-to-date financial information about trade spending, enabling them to easily analyze the company's current liability for trade. Administrators can easily reconcile and clear deductions without wasting time assembling backup documentation and chasing down sales people. Manufacturing gets a clear understanding of upcoming promotional activities to help make more informed decisions about production and distribution.

What is SaaS TPM?

Software as a Service, or SaaS, is the hosted delivery of software, meaning customers subscribe to the solution and access it over the Internet rather than buying and installing a piece of software. SaaS is much more cost effective because it reduces IT burdens and implementation times, and eliminates manual upgrades.

What should I look for in a TPM Solution?

Look for the following:

- An application designed for simplicity for the end user

- Ability to plan at multiple product and market levels

- Highly intuitive application

- Relieving the need for an initial involved training session, or reoccurring training for new employees

- Encompassing the entire trade plan, from AOP to settlement and analysis.

Is TPO Different from TPM?

In a word, yes. You can think of TPM as the heavy lifting involved with the day-to-day operations of planning, forecasting, executing, and reconciling promotions. Trade Promotion Optimization, or TPO, would complement TPM. It uses advanced modeling and predictive analytics to suggest optimized objectives, tactics, or pricing – essentially the coupling of consumption data to shipment data, which allows you to better understand pricing and activity in relation to the category. A good foundation of promotional data from a solid TPM system is a prerequisite to TPO.

CASE STUDY: Bright Idea for Sunny Delight

After spinning off from consumer goods giant Proctor & Gamble, the Sunny Delight Beverages Co. found itself to be a new small-to-medium-sized provider of juice-based drinks. Sunny D executives urgently needed to implement new technologies. They needed to use them in a way that could exploit their position as a smaller manufacturer and help them to compete better with their much larger competitors.

Sunny D executives realized that Excel spreadsheets were not the best way to communicate trade promotion and/or sales and reporting activities between the company and its brokers. Once they began their search, they recognized there was a lot more to trade promotion management than improving and tracking overall trade spend. So they quickly shifted their overarching goal to improving customer profitability.

Sunny D implemented a cloud-based SaaS solution from TradeInsight. Since the implementation, the company has seen a higher level of efficiencies across the board, and has enlisted the help of various departments to help strengthen profits. In fact, the trade promotion management application has already helped the company exceed initial goals for the implementation.

Being able to see trade spend and revenue per customer enabled the groups to perform an accurate analysis of the situation. This resulted in a win for Sunny D and the customer. The TPM solution also helped in other ways. For example, the company sells to the military though government-approved distributors; consequently, they wouldn't normally be able to track the end customer's profitability. By leveraging the functionality of the TPM solution, Sunny D is able to drill down and see how the distributors are loading their data, which in turn enables the company to gain valuable insight into how the end user is buying the products.

Some of the product lines were considered unprofitable. However, by using the system Sunny D was able to analyze the "spend per pack size" data and determine that trade spend was a lot more efficient than initially realized. "The ability to see the details enables us to react faster to market conditions and make the best choices for the company," executives concluded.

Sunny D's sales force is also using the TPM application. In fact, the sales team uses contribution to maximize the efficient use of their customer's trade funding by analyzing top-line revenue down to cost of goods sold. The TPM solution aided in the due diligence process, for multiple product acquisitions, and partnerships.

For Sunny D, it was a bright end to a cloudy beginning.

CASE STUDY: Sweet Solution for American Licorice

American Licorice markets seven brands and over 200 SKUs. The famed Red Vine candy maker used to handle trade promotions the old-fashioned way: in-house, using JD Edwards – and a lot of paper. With so many products and brokers, just keeping tabs on what promotions were running where resulted in little more than a trailer filled with overstuffed banker boxes. Trying to reconcile a deduction meant the sales team had to manually search through volumes of paper to verify a claim.

The company decided it was time to implement a trade promotion management solution from TradeInsight to provide one convenient place to clear deductions, manage promotions, view trade spending, and run reports off funds and markets so it could maximize efficiency and improve transparency throughout the entire company.

Amy Drake, financial analyst for American Licorice, recalls, "We were all really excited on the business side and just a little hesitant on the IT side. We started initially with deductions; then we used the system for checks, and then the sales, and the marketing team tracked trade spending. We began noticing improvements within six months as the technology provided much-needed transparency to the business."

Today, the team at American Licorice has embraced the system across the board, which has brought several company-wide improvements:

- Increase the transparency of sales and spending

- Easily view a latest estimate of spending to see if sales are on track to meet objectives

- Effectively work and communicate with its network of 30 candy brokers and 100 brokerage users to use the trade promotion system and stay all on the same plan, which helps the sales analysts clear deductions and process checks

- Access an accurate view of sales and spend performance quickly and understand how effectively trade spend is being used

- Partner better with customers

- Improve company-wide technological efficiency

- Find deeper insights and involvement with analysts

- Have more visibility into whether forecasts and objectives are driving profitability.

Bottom line: implementing a TPM solution was sweet.

Lorne Schwartz is President and CEO of TradeInsight, a global provider of Trade Promotion Management (TPM) solutions that give CG manufacturers the power to integrate trade promotion management into existing sales and marketing accounting systems to better plan, forecast, reconcile and track trade spend performance across the entire supply chain and gain visibility into spending. For more information, visit www.TradeInsight.com.

CHAPTER 9

The Trend Behind the Spend

By Carrie Shea

The Trend Behind the Spend report reveals insights into today's consumer packaged goods (CPG) trade promotion and shopper marketing spending environment, including a look into the future and how key tactics are changing the retail landscape. AMG Strategic Advisors studied and assessed the trade spending of more than 200 CPG company clients to create this critical and comprehensive report. The full findings include 750+ unique response sets on topics ranging from pricing analysis and slotting tactics to spending priorities and retailer performance. The results include actionable steps for maximizing marketing budgets, gaining competitive advantage and exploiting trends. This report is:

- One of the most robust reports on CPG trade promotion and merchandising spending

- 235 CPG companies represented, across 110 store categories at 55 retailers

- More than $5 billion in trade promotion and shopper marketing spending represented.

The report data was analyzed in two overall categories: (1) Size of the CPG company based on total annual revenue, and (2) Department within the store.

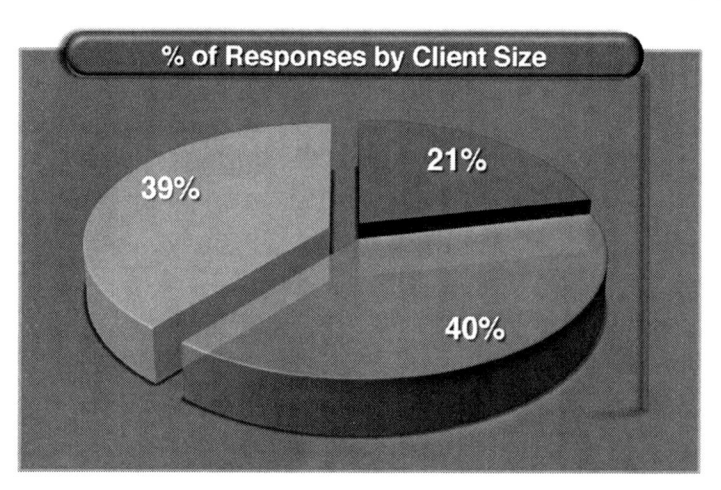

% of Responses by Client Size

- 21%
- 39%
- 40%

■ Small (< $200 Million) ■ Medium ($200M to $1B) ■ Large (> $1 Billion)

Introduction

Tough economic conditions have spurred trade promotion and merchandising spending decisions over the past couple of years. The result has been tighter marketing budgets and increased pressure for quantifiable return-on-investment. The challenge has led to some advantageous outcomes that are expected to continue to accelerate and expand in the world of CPG marketing:

- Better alignment between CPG companies and retailers

- Greater emphasis on shopper marketing strategies.

The largest CPG companies we evaluated are the most likely to employ measurement tactics and analytics to make marketing spending decisions. The missed opportunity for small- to medium-sized companies is significant. The more modest the marketing dollars, the greater the need to be precise and calculating in the competitive world of marketing.

For larger CPG companies, silos among sales, marketing, and manufacturing departments remain a major impediment for effective, results-driven trade and promotion programs. The greatest opportunity for these companies exists in consistently creating and deploying targeted cross-platform, shopper-centric marketing programs.

Going Digital to Win in the Store
Three out of five companies executed digital shopper marketing programs, dominated by Facebook and social media, in the past year. More than half expect to expand digital shopper marketing programs in the coming year. Following Facebook and social media-generated campaigns are Twitter feeds, digital coupons, loyalty programs, and product websites. The challenge is creating cohesive, integrated shopper marketing programs that impact shoppers' behavior and fully understand the ROI on this investment.

Retailer Specific Programs
As retailers have become more aggressive and sophisticated marketers, account-specific marketing has become a more important part of the marketing programs for savvy CPG companies. The larger CPG companies have caught on – some 80% employed account-specific marketing in the past year, but small- to medium-sized CPG companies have been slower to the field. Conditions are ripe for more companies to take advantage of account-specific promotions in the next year.

Key Findings of CPG Companies

Trade Funds
On average, CPG companies spend 13.7% on trade funds as a percent of gross sales. This number is higher for dairy and frozen food companies, and lower for health beauty care/general merchandise (HBC/GM) companies. More than half of the companies using shopper marketing indicated they plan to increase shopper marketing activities in the coming year. Over 80% of large companies use accrual funding vs. only 42% of smaller companies.

Promotion Management

Only 64% of companies analyze most or all of their promotional spending, while two-thirds of companies employ some kind of Trade Promotion Management (TPM) system, and less than 20% have a Trade Promotion Optimization (TPO) system in place. Neither system is likely to see significant increased adoption in the near term.

Pricing

Nearly 80% of CPG companies took a price increase on at least some of their products last year. Deli, meat and bakery manufacturers are most likely to see continued increases this coming year. More than half of the companies surveyed said they are not planning on offsetting the impact to the consumer with trade spending.

Account-Specific Marketing

While two out of three companies engaged in retailer-based account-specific marketing, the spending is markedly heavier for large companies. In fact, more than 80% of larger companies participate in account-specific marketing, while less than half of small manufacturers utilize account-specific marketing. About a third of companies not participating said they plan to implement the technique next year.

Shopper Marketing

Three out of five companies used some form of shopper marketing this year, and more than half expect to increase their usage next year. This number is lower for smaller companies with less than 40% using a shopper marketing strategy. Most of the shopper marketing occurred among the top performing retailers, and CPG companies said they plan to increase shopper marketing activity with these retailers.

More than half of companies reportedly utilizing shopper marketing today plan to increase their activity level in these programs in the future.

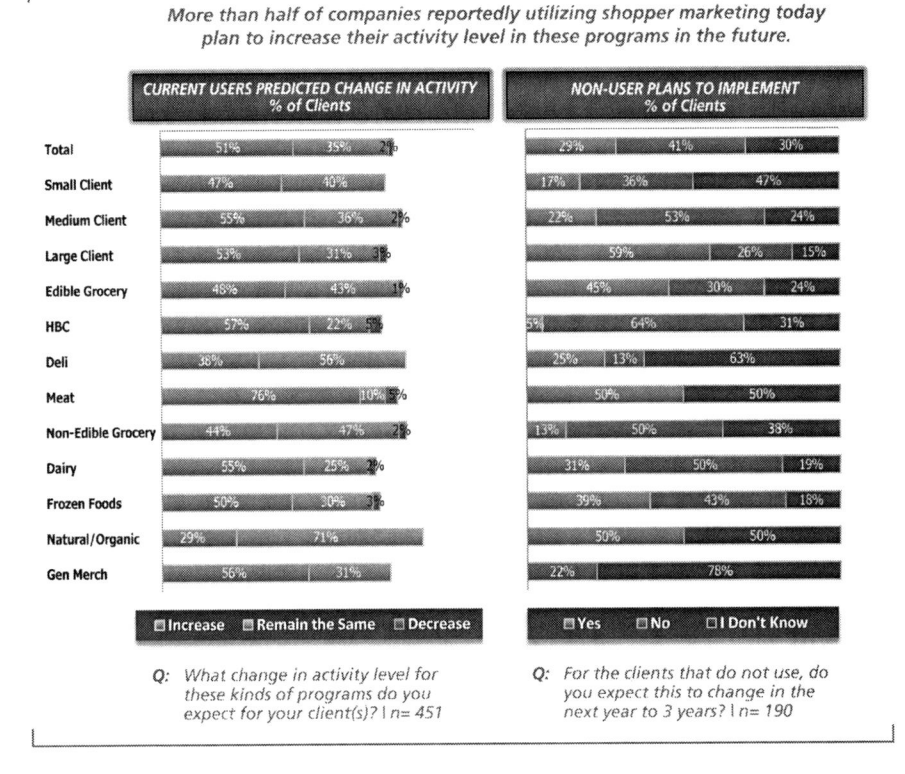

	CURRENT USERS PREDICTED CHANGE IN ACTIVITY % of Clients (Increase / Remain the Same / Decrease)	NON-USER PLANS TO IMPLEMENT % of Clients (Yes / No / I Don't Know)
Total	51% / 35% / 2%	29% / 41% / 30%
Small Client	47% / 40% / —	17% / 36% / 47%
Medium Client	55% / 36% / 2%	22% / 53% / 24%
Large Client	53% / 31% / 3%	59% / 26% / 15%
Edible Grocery	48% / 43% / 1%	45% / 30% / 24%
HBC	57% / 22% / 5%	5% / 64% / 31%
Deli	38% / 56% / —	25% / 13% / 63%
Meat	76% / 10% / 5%	50% / — / 50%
Non-Edible Grocery	44% / 47% / 2%	13% / 50% / 33%
Dairy	55% / 25% / 2%	31% / 50% / 19%
Frozen Foods	50% / 30% / 3%	39% / 43% / 18%
Natural/Organic	29% / 71% / —	50% / — / 50%
Gen Merch	56% / 31% / —	22% / 78% / —

Q: What change in activity level for these kinds of programs do you expect for your client(s)? | n= 451

Q: For the clients that do not use, do you expect this to change in the next year to 3 years? | n= 190

Slotting Allowances

The average slotting cost per SKU varies based on the department and even the size of the CPG company. Slotting applied on a per-store basis tends to cost the CPG company more, but can be beneficial if a store priority/cluster is being used. Paying via free goods yields the lowest average cost.

Slotting Tactics

About 70% of CPG companies are able to negotiate slotting, but less than 30% do so regularly. Tactics are most commonly trade reinvestment or swapping out lower performing SKUs, but other strategies have also proven successful. While 20% overall expect increased slotting fees, dairy and frozen foods manufacturers are most at risk.

> Trade reinvestment and swapping out lower performing SKUs are the most commonly successful tactics when negotiating reduced slotting, followed by bundling products.

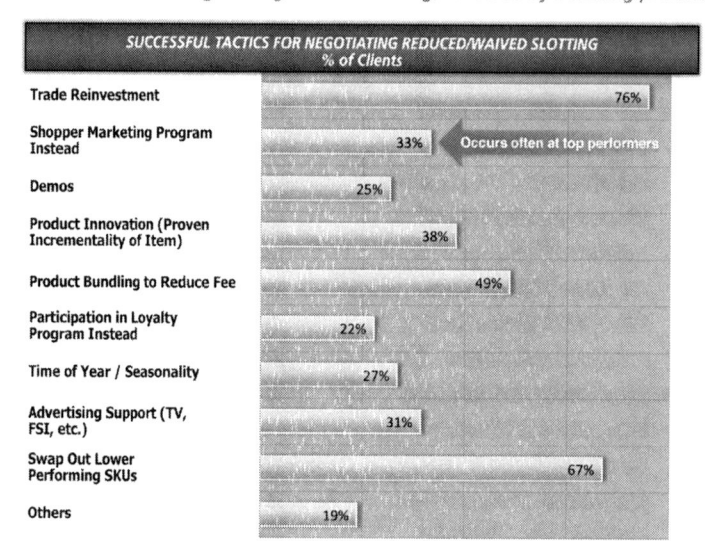

SUCCESSFUL TACTICS FOR NEGOTIATING REDUCED/WAIVED SLOTTING
% of Clients

Tactic	%
Trade Reinvestment	76%
Shopper Marketing Program Instead	33% (Occurs often at top performers)
Demos	25%
Product Innovation (Proven Incrementality of Item)	38%
Product Bundling to Reduce Fee	49%
Participation in Loyalty Program Instead	22%
Time of Year / Seasonality	27%
Advertising Support (TV, FSI, etc.)	31%
Swap Out Lower Performing SKUs	67%
Others	19%

Q: You mentioned that slotting is negotiable at your customer. What tactics have been successful in reducing or waiving slotting for the client? | n= 411

Spending Priority

Based on current economic challenges, the fragmentation of spending choices and the emergence of new digital and shopper marketing opportunities, understanding and prioritizing advertising and promotional spending is more important than ever:

- Top performers grew 4.5% on average; bottom declined 5.0% on average

- Top gained 1.9 points on average; bottom lost 3.0 points on average.

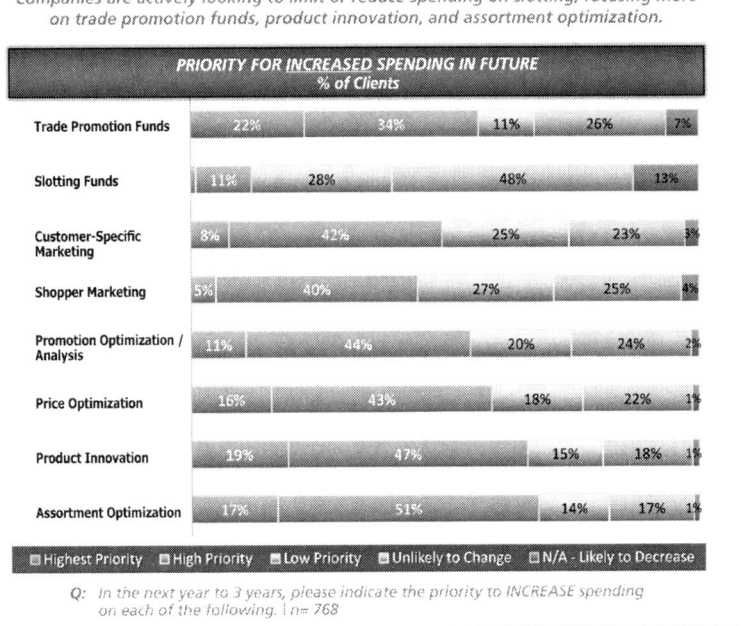

Companies are actively looking to limit or reduce spending on slotting, focusing more on trade promotion funds, product innovation, and assortment optimization.

PRIORITY FOR INCREASED SPENDING IN FUTURE
% of Clients

Category	Highest Priority	High Priority	Low Priority	Unlikely to Change	N/A - Likely to Decrease
Trade Promotion Funds	22%	34%	11%	26%	7%
Slotting Funds	11%	28%	48%		13%
Customer-Specific Marketing	8%	42%	25%	23%	3%
Shopper Marketing	5%	40%	27%	25%	4%
Promotion Optimization / Analysis	11%	44%	20%	24%	2%
Price Optimization	16%	43%	18%	22%	1%
Product Innovation	19%	47%	15%	18%	1%
Assortment Optimization	17%	51%	14%	17%	1%

Q: In the next year to 3 years, please indicate the priority to INCREASE spending on each of the following. | n= 768

The Full Report

This comprehensive AMG Strategic Advisors study was conducted across 110 store categories at 55 retailers with 235 CPG companies represented. More than $5 billion in spending is represented, and unique response sets number more than 750. More than just raw data, the report also includes analysis to identify best-in-class retailers and the spending strategies employed by top accounts. Contact Colin Stewart, cstewart@acosta.com for a copy of the full *The Trend Behind the Spend: A Study of Trade Promotion and Merchandising Spending in the Consumer Packaged Goods (CPG) Industry.*

⌒

Carrie Shea is president of AMG Strategic Advisors. She has more than 20 years experience providing strategy and organizational advisory services to Fortune 500 consumer products, retail, and manufacturing clients. She can be reached at CLShea@acosta.com

CHAPTER 10

Clearing Up Confusion about the Demand Signal Repository

By Janet Dorenkott

There is a lot of confusion in the marketplace regarding the true definition of a Demand Signal Repository, sometimes referred to as Demand Sensing Repository. A DSR is not simply a reporting tool. It is also not simply a database full of demand level/point of sale (POS) data.

A DSR is the process of integrating and cleansing demand data with internal data to provide business users with alerts and reports that will pinpoint areas of the business that require immediate attention.

The acronym, DSR, was coined in 2004 by Lora Cecere and Kara Romonaw, who are experts in the consumer packaged goods (CPG) industry. They defined DSR as "a robust, centralized database that stores large volumes of demand data, such as POS data, inventory movement, promotional data, and customer loyalty data, to support the near real-time decisions that support demand-driven supply networks." The definition continues to evolve today.

The DSR terminology was quickly accepted by the industry and adopted by vendors. Analysts began lumping every software company that could create a sales report into the DSR space. This caused a lot of confusion in the market. Unfortunately, the confusion continues today.

To begin distinguishing between some of the solutions on the market, I began referring to our solution as an *Enterprise DSR*. Coming from the architecture side of business, it was apparent to us that solutions on the market varied greatly based on the long-term manageability and flexibility of the application as well as information provided. We began differentiating the *Enterprise DSR* versus *Team* or *Departmental DSR* offerings as a way to explain the differences. These have become commonly accepted terms and have helped to clear up some of the confusion. Unfortunately, many of the vendors who do not understand architectures refer to themselves as Enterprise DSR providers. Later in this chapter, you'll find some helpful hints to help differentiate solutions for yourself.

Demand data typically refers to point of sale (POS) scanned data. This data is collected by the retailers and can be sent to CPG manufacturers or via third-party data providers. Data maybe sent monthly, weekly or daily. Retailers sometimes charge manufacturers for the data. This varies based on the retailer, the vendors, status with the retailer, etc.

Sources include data coming directly from retailers via EDI, AS2, csv, txt or via retailer portals such as RetailLink, Partners On-Line, etc. Additionally, data may come from data providers such as Nielsen, Symphony/IRI, NPD, Spectra, TDLinx and others. Data providers can supply a variety of information. They can provide raw sales data, reports related to sales or other data such as panel data, market information, etc. Companies may also be getting data from other parties such as brokers and distributors. Other third-party data can include more specialized information such as weather trends, currency information, etc. In an Enterprise DSR, companies will want a streamlined and manageable way to integrate POS data with third-party data and internal data such as shipments, budgets, forecasts, promotions, etc.

The type of DSR you are looking for depends on your needs. If you are a category captain whose need is to create Monday morning reports for your buyer, a Team/Departmental DSR maybe all you need. But if your needs call for supporting many departments that can find value in leveraging POS data, then an Enterprise solution is required.

A solution that is built from the ground up as a reporting tool for a specific retailer is a Team DSR. As stated above, many Team DSR vendors don't understand the differences. Many of these solutions were developed by people who used to work for a retailer and thus developed an easier way to create buyer reports. Wal-Mart solutions are the most common and there are several Wal-Mart reporting tools that now try to position themselves as Enterprise solutions. To a certain extent, they may even believe it themselves because they may have figured out how to integrate additional sources of data. This is typically done with a third-party extraction, transformation, load tool, or via code. Although they may know how to integrate data, what they don't understand is that it's *how* they are doing it that differentiates a true Enterprise DSR and a warmed-over Team DSR. The process, manageability, flexibility and efficiency with which the solution integrates other data sources is what will truly differentiate these solutions.

An Enterprise DSR is not simply a product or database that stores data. It is a framework, a methodology and a process. Its framework streamlines the process of integrating and cleansing POS data with internal and third-party data to give companies a complete view of their sales and business. It provides business users with alerts and reports that will pinpoint areas of the business that require immediate attention. It offers the flexibility to support other applications that could leverage POS data. It has the management tools that allow the application to accommodate changes quickly, notify users of issues, and give them the ability to make adjustments.

The goal of an Enterprise DSR is to provide faster access to more information, improve retailer relationships, maximum ROI, and streamline internal efficiencies. It should make category management easier, reduce out of stocks, improve supply chain visibility, make you smarter about your business, support many different teams and departments and have been built from the ground up to do so.

The Enterprise DSR has an open architecture. This means it will be able to source virtually any database or data type. The target should be capable of residing on multiple database platforms. It should allow end users to work with whatever business intelligence tool they are comfortable with. It should also offer a powerful tool itself with actionable reports. The process should include the integration with internal master data, data cleansing, cross referencing, synchronization, integration with other data sources and loading of the data. It should also have easy-to-use management and administration tools.

The Enterprise DSR should have a data model designed for growth and change. The needs of companies as well as business users are constantly evolving. Retailers and the data they provide will also continue to evolve. Ensuring you have a manageable solution is necessary if you are selecting an enterprise DSR that will be around for years. Consumer goods companies are especially vulnerable to changes. Changes that could affect the DSR include data coming from any retailer, new data in your internal systems and third-party data you may want to leverage. In addition, there may be new third-party data sources that you want to leverage in the future that you are unaware of today. Sources such as social media chatter, loyalty data and web sales are a few examples that most companies never even considered ten years ago.

Team DSR solutions do not have an enterprise integration engine. Companies trying to use coded solutions are especially vulnerable to change. Code needs to be managed every time changes are made to the retailer's data. It means anytime a retailer changes their data feeds, someone has to go in and change the code to accommodate those needs. Changes need to be made to code if your company implements a new ERP system, forecasting application, trade promotion application, etc. If your company merges with another company, code must be

Clearing Up Confusion about the Demand Signal Repository | 137

changed. If two retailers merge, you need to mange those changes in the code. If a retailer starts sending new data, coding is needed.

Anytime anything changes on the source or target or something new needs to be integrated, someone has to physically go in, find where the code changes need to occur, and write those changes into the code. Hopefully, they do it correctly. You also have to hope the vendor you selected provides you with the ability to see and change their code. If a vendor won't give you access to make changes to the integration and cleansing processes, you are working with a Team solution and one that will likely leave you dependent on that vendor.

Team solutions, retro-fitted to be Enterprise solutions, are difficult to manage, time intensive, and costly to maintain. Eventually if a Team solution is incorrectly selected as an Enterprise solution, it becomes a house of cards that will collapse under its own weight. Team solutions should design reports for the retailers they were designed for and not try to be something they are not.

An Enterprise DSR will leverage an integration engine. An Enterprise provider will show you the integration engine including jobs designed to cleanse and integrate multiple retailers. They will show you your management console that will allow you to manage your data feeds. They will show you how you can easily grow and maintain the application yourself or allow the vendor to do so. An Enterprise solution will have a nice tool, but it also allows other tools to query it. Enterprise solutions are open and don't force you to use a proprietary tool. They will support multiple databases including both relational and mpp (massively parallel process) platforms.

If what you need are retailer reports, a Team DSR may be all you need. But if corporate is looking for a solution that will not only provide the retail reports, but also generate cross-retailer reports for management as well as provide reports that will support supply chain teams, category managers, marketing, forecasting and other teams down the road, you need an Enterprise DSR.

If an Enterprise DSR is the goal, include your IT Department in the process. It is critical to understand architecture. Avoid the pitfall of evaluating solely on reports. Any vendor can create reports with data from multiple sources. The data in the report is only as good as the data you are reporting from. POS data is notorious for having data issues.

When evaluating DSR options, I recommend categorizing the vendors into four areas: Enterprise DSR, Team/Departmental DSR, Data Provider or Specialized Application. I add the Specialized Application because we are even seeing Trade Promotion Optimization (TPO), Trade Promotion Management (TPM) and Forecasting applications in the market claiming to have a DSR "built in."

Consider the following questions: What is the Vendor's history? Was the Application designed for specific retailers? Is their focus to provide data? How is the solution integrating internal data? What do their processes look like? How is the data modeled? How are they integrating master data? How are they supporting hierarchies? What capacity for growth does it offer? How do they handle historical data? How fast can data be loaded? How are new retailers integrated? How are data errors handled? Do you have the option to have the application behind your firewall or in a hosted environment? If so, what functionality might you lose if it's hosted? (Do not confuse a hosted environment with a SaaS environment. SaaS environments let you chose from reports, but they are not customized. Hosted solutions are flexible and basically allow you to implement the application as if it were behind your own firewall). Are you dependent on the vendor to add more retailers? How is the integration of data scheduled and automated? Is it possible for you to edit the data? Can the clean data be used for other applications?

No doubt, the DSR market is confusing. But as the market continues to evolve, vendors, analysts and customers are beginning to recognize the differences. Hopefully, the items listed above provide you with some questions that will get you closer to identifying the differences.

Janet Dorenkott is co-founder and COO of Relational Solutions, Inc. The Westlake, Ohio-based company has developed white papers and other documents that focus on the subject of Enterprise versus Team DSR solutions. For more information, visit www.relationalsolutions.com or contact her at janetd@relationalsolutions.com and 440-899-3296 (ext 25).

CHAPTER 11

Using Virtual Environments to Measure Shopper Behavior

By Cathy Allin, Alex Sodek, and Valla Roth

Technology has transformed the shopping process along the entire path to purchase. With the proliferation of technological tools available, shoppers read product reviews and price-shop up to the final purchase decision moment and are now more than ever proactively deciding what to buy, rather than passively receiving messages from the more traditional media. They have exploding numbers of social media sites, blogs and communities at their fingertips, all day and every day. And retailers are using technology more effectively as well, investing in data acquisition and analytics so the right message can reach the right shopper at the right time.

But with all of the technology-driven changes to reach consumers – and shoppers using technology to their benefit – there is one touchpoint that hasn't changed: the bricks and mortar retail store. In fact, it's been said that more people visit Walmart each day than watch the evening news. It's in the store and at the shelf where the buying decision is most often made – in the context of competing brands, on-shelf communications, price reductions, displays, and a whole host of other stimuli that can induce or distract from the purchase.

Technology can be effectively used to better understand the shopper's decision-making at the shelf. It allows you to test and validate ahead of time which strategy – such as shelf placement, price point, packaging graphics, or in-aisle signage – is most effective so that the shopper puts *your* brand in his or her shopping cart.

This technology is virtual shopping research. Manufacturers and retailers use online virtual environments to test whether a shopper marketing program will be effective in the store. Here are some typical questions that virtual shopping research can address:

- Which brand bundle on display generates the most incremental volume?

- Is the new shelf arrangement optimized?

- Should the new line extensions be placed with the parent brand or in a destination with similar products? How does it affect conversion?

- A key retailer is planning to de-list our brand. How can we convince them they should keep it?

- Are the new packaging graphics superior to the ones currently on the shelf?

- Will the new pricing strategy be effective? How will that affect purchasing of our brand? Will we lose volume? To whom?

- The retailer is proposing to place their private brand side by side with our brand. How can we leverage our brand equity and maintain sales for the category?

Virtual shopping research is effective because it is based on observing shopper behavior rather than asking shoppers what they would do. *People often do not do what they say they would do.* This behaviorally-based approach relies on providing the appropriate context (the retail store) so the results replicate what happens in market. Validation studies prove it: virtual shopping shares correlate very highly with in-market shares. Because of this, many retailers readily accept virtual shopping results and implement new programs without needing an in-store test.

How Virtual Shopping Research Works

A virtual shopping study has four steps:

1. *Pick the appropriate shoppers*. It may be based on category usage, brand usage, the retailer's segmentation model, demographics, regionality, etc. – or some combination of them. Many virtual shopping studies are conducted online, so shoppers are recruited via the internet to participate in the study on their own computer and access the survey via a URL link.

2. *Recreate the retail environment*. Set the appropriate context by putting them in the buying mode in the channel or specific retailer. Respondents are introduced to the shopping exercise with digital imagery taking the shopper from the store's parking

lot into the store and to the appropriate aisle, where they shop as they would in real life.

3. ***Measure the sales impact.*** Allow the shoppers to navigate in the store and interact with the products as if they were actually shopping. Provide the experience of picking up a product, looking at all parts of the package, and deciding how many of each product, if any, to put in their baskets.

4. ***Understand the whys behind the shopping behavior.*** This is an important step to provide key reasons to sell the program into the retailer and to optimize the program, if needed. There are a variety of diagnostic tools that leverage technology to provide a better understanding of the offering:

    - **Findability**: Can your brand be quickly and easily found when the aisle has been rearranged or the package graphics modified?

    - **Focus Point** ™: What elements of the package (or aisle) do shoppers see first? What do they see next? We can then infer how their eyes move across the package.

    - **Hot Spot**: What areas of a package (or aisle) do shoppers find appealing, unappealing or confusing? Why?

    - **Virtual Ethnography**: Shoppers in the study participate in online chats with trained moderators to understand more deeply why they made the shopping choices they did.

    - **Eye Tracking**: A virtual shopping study can be combined with eye tracking to measure eye movement across the aisle or individual packages.

    - **Neuroscience**: A virtual shopping study can be combined with a neurological study to better understand how shoppers subconsciously react to the aisle, signage or packages.

An additional step often utilized is called virtual visualization. The resulting data from a virtual shopping study is synthesized into the key insights to present to retailers in a short video. This informs and educates the retailer of the winning new program so that they see the benefits of activating it. In recent years, retailers have become more confident in virtual shopping findings and are resetting their shelves and activating marketing programs based on that information

Therefore, virtual shopping research utilizes technology to create a win/win/win. It identifies Shopper Marketing programs that are a win for the manufacturer because their brand benefits. Well constructed strategies also benefit the retailer by growing the category and creating stronger brand loyalty to their store. And, most importantly, they are a win for the shopper by surprising and delighting him/her with an easier, more rewarding shopping experience.

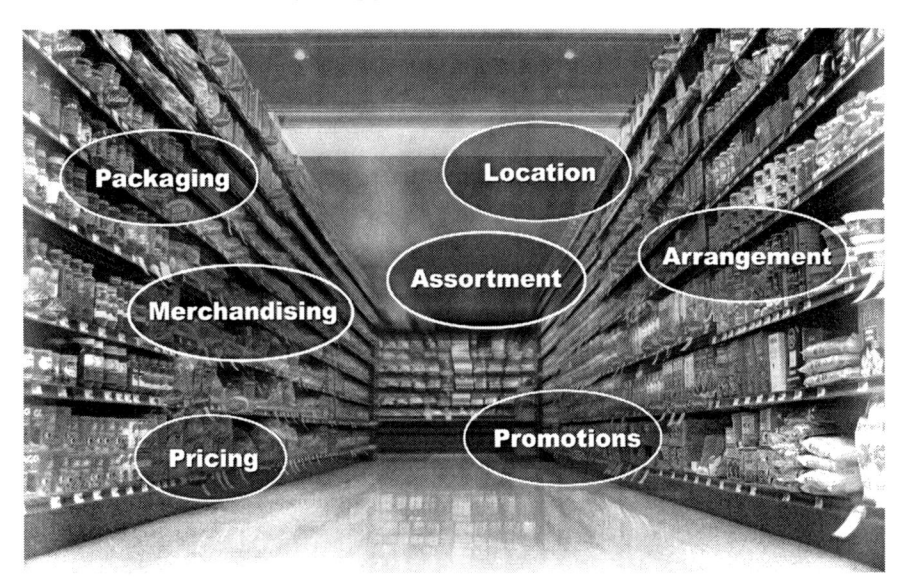

Virtual Shopping Applications
Below are five effective scenarios for which virtual shopping technology provides valuable learning to develop winning shopper marketing strategies. *(Also see the two full case studies at the end of this chapter outlining work with Nestlé and Barilla.)*

1. Arrangement/Assortment. The right mix of appropriately arranged products positively impacts category sales and shopper satisfaction. First, the right assortment of products for the right target shoppers at each retailer needs to be determined. Then the optimal arrangement and flow can be put in place. A leading manufacturer in the category often partners with the retailer to create several alternative planograms based on how shoppers view the category. Virtual shelf sets are made from the planograms and tested among shoppers in a virtual shopping study.

Cadbury utilized virtual shopping technology to determine that a vertical cough drops shelf set (brands and private label arranged in vertical blocks) was much more effective than the existing checkerboard layout (private label SKUs adjacent to branded SKUs). The new planogram grew the category as well as sales of HALLS and private label cough drops, providing the impetus for retailers to reset the section. The results aligned very closely with in-market data. "Vertical brand blocking made it easier to see the store brand cough drops and because the new layout made shopping seem quicker, shoppers perceptions of the store increased as well," said Cadbury executives.

2. Alternative Locations. When a new brand or flanker enters the market, shelf placement is not always apparent. Should a flanker be next to the parent brand to take advantage of the equity umbrella, or should it be with other competitive items in a sub-section? The answer can vary by category and even by retailer. It's especially important to test in large categories where it's easy for new offerings to get lost.

By using an online virtual environment to replicate grocery frozen foods departments, Nestlé learned that ice cream cups should be placed together in their own dedicated area rather than adjacent to their parent brands. This learning prompted the vast majority of retailers to create an ice cream cups sub-section, resulting in incremental space and strong cup sales growth.

3. Pricing. Having the right price not only affects the bottom line, but positions the brand to the shopper in the context of competitive brands. Virtual shopping can be used to gauge customer reaction to price increases before they are implemented. What-if scenarios can be created where competitors do or do not follow the price increase, something that can't be *a priori* tested in the market. Alternatives can be created where both the pack size and price are simultaneously varied in different ways (a WUPU strategy) to optimize the size/price relationship. ROI on promoted pricing options can be determined, such as evaluating various discount levels or different multiple purchasing strategies.

4. Packaging. A package can be thought of as on-shelf advertising that is on the air 24 hours a day, every day. It's important that it works hard for the brand, especially in the context of competitors that are only inches away. Virtual shopping determines whether a packaging change will affect the brand's sales, the most important measure of success. It also provides feedback on whether a new package breaks through the clutter, can be easily found at the shelf, and communicates the brand's messages to increase brand equity, all in a real-world competitive context shelf setting.

5. *Merchandising and Promotions.* In-store communications such as circulars, temporary price reductions, bundled displays, and shelf talkers can have a significant effect on short-term volume. Displays with different combinations of brands, varying unifying messages, and secondary locations can lead to a wide range of incremental sales. Virtually testing several alternatives prior to implementation allows for experimenting with more "out of the box" programs that may be home runs in the market, but too risky to try without shopper feedback.

Why Virtual Shopping Is Better

Technology has allowed the transformation of testing and validating shopper marketing programs in physical stores to testing them in virtual stores. In fact, virtual shopping studies have many advantages over in-market testing. Here is why:

- Virtual tests are more controlled. No worrying that the shelves weren't restocked or that a competitor bought all of the product off the shelf.

- Physical products aren't necessary. If the test is for a new product, the plant may not be ready to produce the product in time for the pre-launch test.

- Virtual shopping studies are quicker. Results are delivered in weeks, not months.

- They are much less expensive, allowing a variety of what-if scenarios to be evaluated.

Virtual shopping studies are also more effective than relying on historical data or informed judgment as to which program will work best. Even sophisticated approaches such as marketing mix modeling rely on the past, which may not be a good indicator of shoppers' future behavior. Using history to generate alternative approaches (such as alternative planograms, pricing scenarios and display alternatives) is a good start. Testing them within a competitive context using virtual shopping, with an eye to the future, is the appropriate next step.

Using Virtual Shopping the Right Way

Whether you have yet to conduct your first virtual shopping study or you are very experienced with them, here are critical components to consider as you design a study:

Represent the channel or retailer. One size doesn't fit all. Shoppers' mindsets and purchasing behavior can widely differ by channel or retailer. Choice sets, aisle layouts and price points vary by venue and can have a large impact, especially in categories with strong private label and regional brand presence. For example, a well known analgesics manufacturer wanted to assess the viability of introducing a flanker brand. To make an informed decision, they conducted the online virtual testing in Walmart, Target, CVS, Walgreen's and Kroger virtual stores.

Create the right competitive context. Large categories such as ready-to-eat cereal, ice cream or carbonated soft drinks have hundreds of SKUs. While they can be accommodated in many virtual shopping platforms, are they all necessary? It depends on the objective of the study. If the issue is at the category level, such as assortment and arrangement, having the shopper interact with the entire category is imperative. Measuring category growth is a key deliverable. If the objective is at the brand level, such as a packaging change, then showing the entire category may not always be necessary.

Target the appropriate shopper segments. Who you test among is just as important as what you show them or what information you collect. Respondents must reflect the population you intend to reach. In the case of online virtual shopping studies, internet usage is so widespread that attaining representative samples is not a concern for most business issues. In most cases you want to screen for category users because most shopper efforts are designed to grow share. In some cases a broader sample is appropriate, such as for a breakthrough product in an emerging category. It's also important to test among shoppers who shop the channel or retailer portrayed in the study.

Beyond the Traditional Retail Environment

Virtual shopping technology can be effectively utilized in venues other than bricks and mortar retail. In the e-commerce space, virtual shopping has morphed into virtual surfing, or understanding how consumers interact with web pages/apps. Traditionally, alternative e-commerce strategies are evaluated using A/B testing in which real-time web traffic is split into varying versions of the web page. However, there are potential pitfalls with real-time A/B testing. First, if an alternative is poor, there is a risk of alienating live consumers. And while it can be determined which alternative generates more traffic, hits, and clicks, there is no understanding of why. Now, a virtual surfing study can be conducted. It begins by setting the context of a natural path to the web page or app being considered. Once at the page, respondents interact with it as they actually would. The study measures how consumers navigate the page, whether they respond to the offer, what elements they notice, and which ones they like and dislike.

Questions to Ask

Having a better understanding of what virtual shopping technology provides, it's time to put it into practice. Below are some questions to ask when selecting a virtual shopping partner:

- Can they provide validation of their research and resulting data across multiple categories?

- Do they have an established track record of providing results leading to activation at retail for clients?

- Can they show it delivers measurable ROI on its research?

- Do they understand shopper marketing issues and interpret and deliver results strategically?

- Do they include rigorous diagnostic tools to understand the *whys* behind the shopping data?

- Is the user interface intuitive and easy to navigate?

- Is the respondent required to download a plug-in or special software to participate in the study? (This can have a negative impact on sampling.)

Summary

Technology has transformed the entire path to purchase by providing many tools for shoppers to research, comparison shop, connect with other product users, and even make the purchase. However, there is one shopper touchpoint that hasn't changed: the bricks and mortar retail store. It's at the store, in the context of competing products, where the purchase decision is most often made.

Retailers have a variety of levers they can pull to influence purchase decisions such as assortment, arrangement, in-aisle promotions, pricing and packaging.

Technology can also be used to better understand shopper decision-making at the shelf; to identify in advance the optimal assortment, arrangement, pricing, packaging or promotion. This technology is virtual shopping. The use of virtual environments has become a proven process among leading manufacturers and retailers to test and validate alternative in-store strategies and shopper marketing programs.

Virtual shopping is an effective tool because it can deliver an accurate representation of in-store performance by observing shopper behavior, rather than relying on asking shoppers what they would do.

Virtual shopping technology provides a win/win/win solution. The manufacturer wins because a new initiative is implemented to grow sales and share. The retailer wins because well designed programs grow the category and build loyalty to the store. Most importantly, the shopper wins by having a more rewarding, easier shopping experience with offerings better meeting his or her needs.

By recreating the shopping experience, virtual shopping technology can be effectively used to answer a variety of strategic and tactical shopper marketing questions. It is a cost effective, timely and validated way to garner insights as well as to provide retailers the evidence they need to activate the desired changes in their stores.

CASE STUDY: Nestlé's Merchandising Location Strategy Wins with Retailers

Nestlé, the world's largest food company whose portfolio includes well known ice cream brands such as Edy's/Dreyer's, Haagen-Dazs and Skinny Cow, wanted to revitalize the slow growth ice cream category. But how does one revitalize a long-established category that already has nearly universal penetration? Nestlé chose to frame the business differently by looking at it in the context of snacking. At $100 billion, the snacks category dwarfs the $10 billion ice cream category, and it is growing. In fact, nearly two in three people snack four or more times a day. This could lead to new usage occasions for ice cream and grow the buy rate for the category.

Nestlé began the exploration by creating a "roadmap" that identifies broad consumer trends and key retailer needs that could lead to growth strategies. According to Russ Onish, former Director of Category Leadership & Shopper Insights for Dreyer's Grand Ice Cream, Nestlé USA, "We decided to focus on three consumer needs: wellness, convenience, and variety seeking. This naturally led us to ice cream cups." Unlike packaged ice cream, single-serve cups offer portion control (wellness), they can be eaten from the cup (convenience), and it's easy to buy a combination of flavors for the household (variety seeking). Not surprisingly, the concept generated strong consumer feedback. At a higher price per serving versus packaged ice cream, there was a great retailer story to trade consumers up to increase profitability.

Nestlé was getting ready to launch a new line of cups for the 2010 season, and planning to make a big impact to the cups category by introducing 15 SKUs. The flavors were identified, formulations were finalized, packaging was created, and production plans were made. But an important retailer question surfaced: now that the cups category will have critical mass, should the cups be shelved together in dedicated doors or should they be part of their respective brand families? The Nestlé brand teams predictably wanted to maintain brand blocking, while the innovation team saw the opportunity to create a new destination with a new eating occasion and different consumer benefits.

One way to answer the question would be to test each configuration in a handful of stores in market. But there wasn't time: important retailer meetings were coming up.

Nestlé turned to a virtual shopping platform to answer the merchandising location question as well as a secondary question on pricing – whether the promoted price should be 99¢ each or 10 for $10. The alternative arrangements were tested in two virtual retail environments: Safeway and Kroger. Category shoppers were recruited online to participate in the study. Video fly-throughs moved shoppers from the parking lot into the store and to the ice cream section. Once in the aisle, they shopped the category as they normally would. The shoppers could buy as much of any product as they wished, or even walk away from the aisle without purchasing anything. It was a large study with 20 freezer doors and over 400 SKUs.

Initially, the project was designed to test two merchandising location scenarios: dedicated cups doors and cups with the parent brand. But as with many business decisions, it became apparent it wasn't that black and white. Skinny Cow is different from the other cup brands. Being a better-for-you product, the brand has a very loyal following that knows exactly where to find it. And once she finds it, she tends not to shop the rest of the aisle. Skinny Cow doesn't have packaged ice cream offerings; it is in the ice cream snacks section. And, it is higher priced. Skinny Cow cup promoted pricing is $1.25, so it couldn't be promoted with the other cups at 99¢ or 10 for $10.

With these considerations in mind, the team decided to have respondents shop totally dedicated cups doors, dedicated cups doors sans Skinny Cow, and cups with the parent brand – all with and without promotion.

Thinking of Skinny Cow as a different animal paid off: the strongest location strategy was to merchandise the cups together, except for Skinny Cow. This scenario created the strongest category revenue, as well as the strongest sales for the Nestlé portfolio. Importantly, it motivated shoppers to variety seek which is a fundamental part of the business plan.

The promoted pricing results point to the 10 for $10 strategy. Slightly more shoppers are attracted to the multiples price point, and they buy more cups at a time, resulting in stronger revenue.

Within weeks of receiving the virtual shopping results, the Nestlé shopper marketing team presented the merchandising story to key retailers. Feedback was positive and plans were made to create the dedicated doors for the 2010 season.

The in-market results are impressive: 90% of stores stock cups together, and cup revenue is up 53% in those stores. Nearly as many stores (85%) followed the recommendation to keep Skinny Cow as a separate brand block, and as a result Skinny Cow sales have more than doubled in those stores.

Overall, ice cream cups are showing explosive growth. Creating the right "on-trend" consumer offering, providing the retailer the right reasons to carry it, and merchandising in the optimal location is clearly a recipe for success.

CASE STUDY: Barilla Conquers the Path to Purchase: Influencing Change at Retail

Innovation had recently occurred in the pasta category with better-for-you options being introduced, such as whole wheat pasta and those with Omega-3 added. Being a category leader, Barilla wanted to determine the best pasta aisle planogram to recommend to retailers. The key question was to understand where to locate the Better-for-You (BFY) pastas: next to each one's parent brand or as its own vertical brand block? The optimal arrangement needed to both grow category dollar sales and increase retailer loyalty.

Barilla chose to test the planograms using an online virtual shopping platform. Category shoppers were recruited online to participate in the study. Video fly-throughs moved shoppers from the parking lot into the store and to the pasta aisle. Once in the aisle, they shopped the category as they normally would by clicking on a product to see product details (as if they were picking it up), and then deciding whether to put it in their shopping basket or not. The shoppers could buy as much of any product as they wished, or even walk away from the aisle without purchasing anything. In addition, trained moderators chatted online with the participants during the course of the study to better understand the "whys" behind their choices and create deeper shopper insights.

The study revealed that placing BFY pastas with the parent brand is optimal for several reasons:

- It increases Barilla revenue.

- It increases total pasta category revenue.

- It attracts heavy pasta users, a very important segment: both BFY awareness and BFY penetration are stronger.

- It strengthens retailer loyalty. Especially among heavy users, shoppers are more likely to return to the store.

Some valuable insights were gleaned from the one-on-one online chats which provided more depth to the learning:

- Pasta is a brand-driven, non-impulse purchase. Shoppers choose a brand based on quality and value, and become loyal to it.

- BFY pastas have an image of not tasting good and being more expensive.

By placing the BFY pasta next to the parent, brand loyalists will see these SKUs next to their favorite pastas. Because of the halo effect of the Barilla brand, they are willing to try the BFY products since they know they are better for their family and the trusted brand name alleviates taste and price concerns. This variety-seeking purchase behavior is even stronger for heavy users.

The study successfully determined which aisle performed best, and why. It delivered valuable insights for Barilla. These results provided Barilla with the information they needed to get the new planogram implemented at key retail partners. In-market results were in line with what the virtual test predicted, indicating it's a win for both the retailer and Barilla.

Cathy Allin is President & CEO of Decision Insight, a provider of virtual shopper research to help understand the impact of in-store strategies at the point where the shopper makes his or her buying decision. Alex Sodek is Chief Research Officer and Valla Roth is Director of Communications for the Kansas City-based company. For more information, visit www.decisioninsight.com or email info@decisioninsight.com.

CHAPTER 12

Understanding 'Ready-for-Purchase'

By Andy Kalamaras

The hand-wringing over retail out-of-stocks is a never-ending challenge. Despite decades of analysis, discussion and attempted point solutions, the statistics don't seem to improve. Across the grocery industry, the 8.3% out-of-stock rate reported by the Grocery Manufacturers Association (Gruen and Corsten, 2007) has been reluctantly labeled as an unavoidable "cost of doing business."

This is not for lack of valiant effort. Solutions such as computer-assisted ordering (CAO), perpetual inventory (PI) and increased store labor have had limited success, but left us with fatigue, skepticism, morbid belief and resigned acceptance. Nevertheless, as the Gartner Group reports, "more than 65% of out-of-stocks [are] attributed to processes within the four walls of the store." (Griswold, 2011)

Providers of point solutions have long maintained that *post hoc* analysis should lead to better merchandising and distribution plans that will drive down out-of-stocks. The solutions are crude. We can increase facings to add safety stock on the shelves, but suffer lower turns and higher inventory carrying costs. We can increase replenishment frequency, however the labor cost can be prohibitive. Experience tells us we can't just plan our way out of the issue of lost sales.

Two promising factors are leading the industry to a new opportunity on this issue. The first is a clearer understanding of what defines an item as "available." The second is a breakthrough technology that taps into a vast source of real-time intelligence about shelf conditions – the shoppers themselves.

Item Availability from the Shoppers' Point of View: 'Ready-for-Purchase'

It's time to re-frame the conversation about how shoppers confront what we call "out-of-stocks." The old way was limited to after-the-fact reports that provided an incomplete picture of what items were not sold, with little insight about why. The new strategy monitors point-of-sale (POS) transactions and applies pattern recognition in real time to maintain a continuous understanding of what items are Ready-For-Purchase (RFP) – and which are not – from the shoppers' point of view.

"On-shelf availability is rapidly becoming the key measure of retail supply chain effectiveness," stated a recent Gartner Group report, *On-Shelf Availability* (Griswold, 2011). "With shoppers becoming more empowered, retailers must focus on improving the on-shelf shopping experience to retain and grow market share."

We submit RFP as a *positive* measure of performance that tells us which items, categories, promotions, stores or vendors are delivering results at the shelf. It encompasses more than the simple presence or absence of an item in the store. To be acceptable to shoppers, an item must also be visible, correctly located, correctly priced, clean, in good condition, fresh and not short-dated or out of date.

These are important refinements compared with the legacy thinking about out-of-stocks: Too often, items are present but not Ready-for-Purchase (nRFP). The solution may be as simple as replenishing stock on a fast-selling soup flavor from the back room – or as subtle as checking egg cartons for short dates. The data indicates that upwards of 50% of nRFP items can be immediately corrected *in the store*, with simple merchandising maintenance.

> ### Five Insights for 'Ready for Purchase'
>
> 1. Ready for Purchase means an item is present, correctly located, correctly priced, clean, in good condition, and fresh-dated – in short, acceptable to the shopper.
>
> 2. Most RFP issues are immediately correctible in the store – if you have means to detect and act on them.
>
> 3. More than 20,000 shoppers a week per store are your front-line reconnaissance force for item availability.
>
> 4. Breakthrough technology now makes detection automatic and alerts easy to generate.
>
> 5. Data storage, bandwidth and processing power have advanced to where there's no longer any technical obstacle.

The 20,000+ Member Store Reconnaissance Force

The obvious question this raises, of course, is "Who has the time or resources to track this down across 45,000 items in our assortment?" While some degree of self-reporting by merchandisers might help, their data provides a momentary snapshot at best. Continual audits of actual shelf conditions by store personnel or third parties are unrealistic – prohibitively expensive and of questionable accuracy. And many of the factors that negatively affect RFP are simply too hard to detect at the shelf, even with frequent store walks.

Retailers have access to a massive, under-utilized resource for rapid detection and response that carefully inspects our stores every day – our shoppers. With an average of more than 20,000 visits per store per week, our entire inventory is being scanned and scrutinized on a continual basis by committed people with a deep, personal stake in Ready for Purchase. And they do it for free.

All that's required is a way to tap into the collective genius of all those shoppers. In fact, the output of their reconnaissance efforts are captured at the POS scanners minute by minute. Each time an item is purchased – or not purchased – there is meaning in the data that we can act upon. If an item is not selling as expected, there may be a reason we can check and fix.

Ready-for-Purchase from the Shopper's Perspective

There are numerous scenarios in which a motivated shopper may be unable to purchase a desired item of merchandise. When a desired product is not on the shelf or otherwise unavailable (about one item out of 11 in a typical supermarket), the shopper is forced to either choose an alternate item, find it at another store, or do without. These represent negative outcomes for the retailer – and negative experiences for the shopper.

The outcomes of these events – lost sales, gross margin reduction, and a loyalty-shaking shopper experience – are always regrettable. Retailers need a reliable method to assure that items are Ready-for-Purchase at the moment of decision.

nRFP Two Ways: At left, a classic out-of-stock, caused by a hot promotional price, limited cooler space and/or inadequate back stock. At right, a missing price tag discourages shopper take-away on a premium store-brand item. Item is present, but *not Ready For Purchase*.

Industry dialog around this issue has traditionally focused on what is called "out-of-stock" – instances where product is assumed to be simply missing from the shelf or display. These are typically measured after the fact by reviewing store conditions, warehouse "scratch" reports and item sales rates. But simple inference is not an effective way to restore items to RFP status. As Gartner analyst Mike Griswold states: "On-shelf availability remains a significant challenge for retailers. Despite the focus on supply chain and inventory management initiatives, improvements in shelf execution still elude most retailers. New research based on interviews with retailers and technology providers shows a changing perspective on the definition of 'in stock,' as well as a growing sense of urgency in fixing the problem."

Some RFP Scenarios

Consider the following common retail scenarios, where an item may be physically present in the store, but not ready for purchase:

- A produce display is full of bananas that are visibly over-ripe. POS scan data shows few or no sales.

- A shelf of yogurt cups looks full and neatly arranged, but many package dates show expiration within two days.

- A hot price for store-brand canned beans has depleted shelf stock. A well-meaning clerk has "faced out" the shelf with adjacent items, masking the issue.

- A popular line of frozen entrees is on display at list price. A competitor advertises the same products on deal, and shoppers are stocking up across the street.

These are just a handful of the everyday issues that impact Ready-For-Purchase in your stores. They are often hard to detect, but once identified, 50-70% can be controlled in the store. Often the fix is simple – restock from the back room; rotate out short-dated or unattractive stock; replace a shelf tag. For more complex fixes – a competitive price change, display set-up or re-order of a depleted item – rapid initiation saves lost sales.

Not Ready for Purchase?

Here are some common root causes of why a product is not Ready for Purchase:

- Item not on shelf (OOS, upstream void, product in the back)

- Item not in store

- Item is hidden from view

- Missing display

- Item present, but tag is missing

- Item present, but wrong price posted

- Item present, but tag shows incorrect unit measures or price/ounce

- Out of date or short-dated item

- Damaged/dirty product

- Perishable ripeness or condition is unacceptable

- Spread /parity of prices (for example, brand versus private label) is incorrectly set

- Competitor advertises lower price

- Competing promotion in same store/category siphons sales

- Posted price does not match promo price in circular

Admittedly, practical tools for continually monitoring Ready-For-Purchase data have only recently become available. Historically the reasons why items sell below their potential have been poorly measured. As the examples above imply, shoppers decide to buy or not based on multiple factors including product condition, price and appearance. Mere presence of the product, while essential, is only the first influencer.

To correct merchandising shortcomings, merchants need to better understand why each item is not selling, or selling below the expected frequency. The sheer number of items in a store presents a level of intricacy that can be successfully addressed with automated *purchase pattern recognition*. RFP assurance is a perpetual challenge and it demands a continuous solution – one that is always on and woven into the operational fabric of the business.

Figure 2. How "Ready-For-Purchase" Beats Tracking "Out-of-Stocks"

Old Way (reduce OOS through planning)	New Way (RFP detect and resolve)
NOMENCLATURE • Legacy Concept: "Out of Stock"	NOMENCLATURE • New Concept: "Ready-For-Purchase"
SOLUTION APPROACH • Attempt to predict/prevent • Rely on merchandisers' innate smarts • Not enough labor to keep up • Try to control through planning refinements	SOLUTION APPROACH • Model expected item demand/sales rates • Continually monitor the transaction log • Rapidly detect discrepancies • Alert merchandisers to check and resolve RFP issues
DETECTION METHOD • OOS reported after the fact, *post hoc* data analysis • Accidental discovery by store personnel • Customer complaint or request	DETECTION METHOD • Harness the inherent genius of shopper behavior • 20,000+ pairs of eyes scan the store each week • Compare purchase data against forecast to detect RFP issues
MERCHANDISING RESPONSE • Re-order or de-list item • Reconciliation with trading partners • Adjust future plans • Revise order quantities	MERCHANDISING RESPONSE • 50-70% are correctible in the store • Alerts communicated rapidly to people in stores • Monitors staff response via self reporting • Data accuracy is "baked in" to the model
TOOLS (REPORT AND REPLAN) • Store checks/audits • Price changes • Promotion plan • Assortment/Distribution plans	TOOLS (DETECT AND RESOLVE) • Track shoppers' collective behavior/intelligence • Track and mine 20,000+ transactions/week/store • Gigantic free labor pool • Automatically alerts merchandisers to check and correct
TECHNOLOGY • $$$ High cost to address • Post-event reporting • Sample-based audit data • Tracks how often we failed, on average	TECHNOLOGY • $ Lower cost solution allows "always on" process • Data crunch power is plentiful, so it's no longer a problem • Continuously models expected item movement • Continuously compares sales rates against the model.

Sustainable Competitive Advantage: Benefits of the RFP Approach

Ready-for-Purchase is both a new orientation and a new methodology. It delivers a new standard of retail performance. Legacy methods for combating out-of-stocks have saddled retailers with persistently lackluster results – especially on promoted items. A fresh approach to managing item availability turns this former "cost of doing business" into a sustainable competitive advantage. Monitoring and maintaining Ready-for-Purchase delivers improved service levels to shoppers, detects difficult-to-find nRFP items and captures sales that would otherwise be lost.

Over the longer term, the sales velocity model provides continually improving inputs for existing planning methods. Assortment, space, price, promotion, and new item introduction plans all perform better, because the forecasts used to develop them are more reliable.

When RFP is consistently high (especially on key value items with high affinity), shoppers are more successful. The retailer gains a deserved reputation for "always having what I want." Shoppers are more likely to return, and stores capture a larger share of wallet, trips and heart.

Andy Kalamaras is Senior Vice President, Sales and Professional Services for KSS Retail, a dunnhumby company. For more information: www.KSSRetail.com.

References:

Gruen, Thomas W., and Corsten, Daniel, A Comprehensive Guide to Retail Out-Of-Stock Reduction in the Fast-Moving Consumer Goods Industry, Procter & Gamble (2007)

Griswold, Mike, Improving On-Shelf Availability for Retail Supply

Chains Requires the Balance of Process and Technology, Gartner Group (May 2011)

Griswold, Mike, Improving On-Shelf Availability: Rapidly Becoming the Key Measure of Retail Supply Chain Effectiveness, Gartner Group (September 2011)

SECTION FIVE
DIGITAL

CHAPTER 13

Achieving and Sustaining Planogram Compliance

By Greg Gates

Let's start by stating what we have all come to know and accept – retailers have a planogram compliance problem. The planogram that is painstakingly developed by category captains and retailer headquarters staff just isn't implemented consistently in stores. Even when it is, errors and short-term fixes creep in within a few days. So why does so much of the talk in the industry still focus on measuring compliance and not improving it?

At some point, the focus must shift to solving the compliance conundrum over the long haul. Sustainable compliance is what retailers should be striving for throughout a planogram's lifecycle to maximize profitability. There are significant advantages to taking a sustainable compliance approach. According to the ISI Sharegroup, the total cost of non-compliance is approximately 1 percent of gross product sales in the industry, which translates to a lost sales opportunity of between $10 billion and $15 billion in the food, drug and mass merchandising channels.

Yet standing in the way are some very difficult-to-control factors that can impact a category's compliance over time. Shoppers, stockers and competing manufacturer reps can all shift items around in a correctly set planogram, quickly driving the shelf out of compliance. As time goes on, this continuous movement puts the shelf further out of compliance to the planogram – often by as much as 10 percent per week.

No Magic Compliance Pill

The hard truth is there is no magic compliance pill. Significant challenges exist throughout the merchandising process, from headquarters to the store, from the store the shelf, and even from shelf to shelf within a category in the store. In our work with retail chains, we are finding four frustrating factors that prevent stores from getting – and keeping – their stores compliant:

- Inaccurate and/or incomplete dimensional data
- Poor store mapping
- Out of stocks – including mis-stocks and face-overs
- Budget constraints.

The interesting thing is that retailers have the resources available to solve the sustainable compliance challenges they face by taking a more systemic approach to recapturing some of those lost sales.

Looking at Data in a New Way

In the planning stage, one of the biggest barriers for category managers at headquarters in creating effective planograms is inaccurate or incomplete dimensional data. Planograms built using wrong product dimensions quickly go awry, resulting in space being left on the shelf unused or items simply not fitting in the assigned shelf space. And once the planogram lands at the store, the staff may take matters into their own hands. Why? Quite simply because the planogram doesn't work for their store. That further complicates the compliance conundrum.

There are several reasons dimensions may be wrong – missing items, poor measurement processes, old measurements, or a lack of internal systems. In addition, we find retail chains that try to generate this data internally are often overwhelmed by the number of new products that hit the shelves each year. According to SymphonyIRI, more than 150,000 new items were introduced by consumer packaged goods (CPG) manufacturers in 2010 – with 96% of those being line extensions.

So effective merchandising starts with accurate data as the foundation for planogram development. But the other complicating factor is creating planogram versions that take into account the wide variety of store and shelf layouts, and incorporate timely feedback from store managers. As versions of the original planogram are created, timely feedback on the unique conditions at each store is critical to ensure the planograms can be implemented correctly.

Between the explosion of new SKU data and planogram versioning challenges, planogram development, even for retailers with the deepest internal resources, can quickly become a nightmare. That's why retailers need to get out of the product dimensions and planogram versioning businesses, and get back to letting merchandising teams focus on the big picture. Expert analysts from third parties can be tapped to build accurate, effective planograms using comprehensive, up-to-date dimensional data, with versions that can be adjusted quickly based on local store feedback. This approach can save retailers up to 60 percent of the time they previously spent developing planograms, so merchandising managers can implement category resets more frequently and better match assortments to consumer preferences – recapturing some of those long lost sales.

From HQ to the Store

Effective planning only accomplishes so much. Even the most effective planogram still runs the risk of being derailed in the transition from headquarters to the store. Why? It comes down to the human factor. Forrester estimates that labor costs are between 10 and 13.5 percent of a store's revenue, even with the ongoing contraction in staffing we've seen throughout the industry. Yet new product introductions in a 1,000-store chain can require store associates to execute *700 million* intricate tasks each year to stay in compliance.

Simply put, new product introductions, promotions and store resets quickly engulf a store's labor budget, making it more and more difficult to bring a store into compliance – let alone keep it in compliance over the long haul. That's why new shelf-edge merchandising solutions – such as visually intuitive shelf strips, reset strips, shelf tags and back tags – are so important in helping store personnel set shelves quickly and accurately while also preventing out of stocks. This visual approach to stocking can reduce lost sales due to out of stocks by as much as 65 percent, and decrease shelf reset and new store opening time by as much as 60 percent.

Sustainable Compliance – A Big Lift
Retailers today know that they have to go beyond the planograms to achieve sustainable compliance in order to win more returning shoppers and achieve a larger share of the market basket. A benchmark study from the National Association of Retail Marketing Services (NARMS) found that that retailers achieving compliance can realize a 7.8 percent increase in annual sales. That translates to $3.8 million for a 200-store chain – and an 8.1 percent lift in profit. While the data in this study is now 10 years old, retailers today are finding they can achieve similar gains by addressing compliance over the long term.

By taking a sustainable compliance approach, retailers can successfully tackle some of their toughest issues – labor costs, improving communications between headquarters and the store, and reducing stocking mishaps at the shelf. They realize that sustainable compliance is possible through more accurate dimensional data and well-designed planogram development and image merchandising solutions. More importantly, they realize it is essential to establishing and extending their competitive advantage.

Greg Gates is the Vice President of Image Merchandising Solutions for Gladson, the leading provider of syndicated consumer packaged goods (CPG) product information and services for manufacturers, retailers, wholesalers and brokers. He can be reached at ggates@gladson.com.

CHAPTER 14

Is Your Brand Winning or Losing in the Coupon and Promotions Game?

By Wade Allen

Coupons and other promotional offers are top of mind these days. It seems we can't go a single day without hearing about another daily deal site or mobile coupon app. Combine these additions with the already established daily deal players and digital coupon vendors of Groupon, LivingSocial, Coupons.com and many others, and you start to get sense of the breadth of the marketplace.

But it is not just an increase in vendors. Customers are seeking discounts now more than ever before and are willing to switch to products and brands that provide the best deal. So what's really fueling the fire behind all of this discount demand? There are five factors that are likely the driving forces behind the discount epidemic.

Economic Slowdown
The proverbial music died in 2007. The financial bubble-burst leaving us all in shock, and the after-effects left most people scrambling to adapt. In such times, the thrifty tend to thrive, and so couponing became a key strategy for consumers looking to stretch their dollars to the limit. Even now, as the economy rebounds, couponing and deal seeking continue to flourish. According to a 2010 study by comScore,

66 percent of consumers said they used coupons in July 2010, compared with 59 percent two years earlier. The percentage of respondents who reported shopping online for deals increased from 24 percent in 2008 to 32 percent in 2010.

Product Parity

With product quality, durability, and availability reaching parity, customers are forced to make purchase decisions on the only remaining factor: price. Though an emotional tie to a brand can still influence a customer's purchase decision, its impact is dramatically reduced when customers are strapped for cash. New research by Ipsos Marketing indicates that 80 percent of global consumers feel store brands are the same as, or better than, national brands at providing a variety of benefits including taste, convenience, and environmental friendliness.

Social Explosion

The mass adoption of social networks like Facebook, Twitter, Linkedin and others has created a digital social audience that has never been seen before. To put this in perspective, keep in mind that Facebook grew from 27 million U.S. users in July 2008 to over 150 million in July 2011. Recent numbers show that Facebook has approximately 800 million users worldwide. That means that if Facebook were a country, it would be the 3rd largest and represents 11% of the world's population!

What's more important than the quantity of users is that those users are consumers who are now engaging brands on social networks. They're also looking for more than just content. Simply put, "Likes," retweets, and mayorships aren't enough any longer; it's about added value, and this is where discounts/promotions are king. According to a study by Merkle, social media users who choose to "Like" a brand on Facebook most often do so to receive exclusive deals and offers, including coupons.

Digitization of Coupons and Promotions

To capture this movement, more and more brands and retailers have made their coupons and deals available online. This has led consumers to embrace digital couponing at a staggering rate. eMarketer predicts that 92.5 million adult Internet users will use online coupons in 2012. Another study by Experian Simmons found that the percentage of households using print coupons has not changed much since 2005, but the percentage of households using digital coupons has risen 83.3 percent.

Smartphone Usage

With smartphone penetration surpassing 50 percent in 2012, consumers are poised to receive mobile coupons in a massive way. Additionally, while mobile coupons only represent a small share of digital coupons today, they have enormous potential to become the couponing medium of choice in the near future. Because mobile coupons influence a consumer in the midst of making a purchase decision, retailers and brands are beginning to experiment with this type of delivery standard. Imagine how great it would be to unlock a 20 percent discount while standing in line at a local restaurant. The best part about it all is that a customer can gain access to said discount with only a quick download of an app or a check-in, which takes only a matter of seconds.

What Are Brands and Retailers to Do?

In today's new landscape, opportunity abounds but companies have to "get in the game" to win. One can only assume that brands and retailers that wait and idly stand by will be left to play catch-up in later years. Here are 3 simple suggestions to take advantage of the promotion hype today:

- **Get Smart on Emerging Technologies.** The ever-changing coupon and promotions landscape can be intimidating and make it difficult to stay "in-the-know." Smart brands and retailers will take the time to investigate new technologies and/or seek guidance from expert promotional agencies and vendors. Not every new "shiny object" will be worth investing in. Make sure to weigh the pros and cons of each new solution and how it will impact your core customer. Remember: Those who understand technology are the best positioned to profit from it.

- ***Update Your Strategy and Align Promotion Budgets.*** Brands and retailers that modify their coupon/promotion strategy and align their budgets to include digital print-at-home, save-to-card, mobile couponing and other emerging promotional tactics will stay ahead of the competition. Being proactive now will allow organizations to avoid future headaches as new promotional mediums and technologies expand and firmly take hold.

- ***Monetize Your Social Footprint.*** Brands and retailers cannot afford to pass up opportunities to engage consumers in social properties anymore. A quick and easy way to monetize your social footprint is to add a coupon/promotion tab to your Facebook page and require users to "Like" your brand. This simple tactic will help drive customers to purchase and give your brand a voice with actively listening consumers. Make sure to use a secure coupon technology on your tab to avoid any coupon fraud. Opt for a solution that won't redirect customers away from your Facebook page and will track coupon metrics and customer leads in real-time.

Future

No one can predict the future, but it's safe to assume that couponing and deals are here to stay. Communication media will evolve, technology will advance, and strategy and tactics will change, but brands and retailers will continue to use coupons and promotions to drive customers to purchase. Doing this the smart way will require brands and retailers to recognize consumer demand, accept and engage in new couponing technology and mediums, and adapt their promotions strategy and budgets.

Now ask yourself: Is your brand winning or losing in the coupon and promotions game?

Wade Allen is president of CouponFactory, provider of a digital coupon platform that allows companies to quickly create, manage, distribute, and measure digital coupons. For more information, go to www.couponfactory.com.

CHAPTER 15

Better, Faster, Smarter Shopping Experiences

By Philippe Loeb

Consider what is happening in the marketplace today:

- As many as 80% of products fail within their first year of launch, according to research by the University of Toronto.

- 70% of all purchase decisions are made in the store, often in a matter of seconds, says the Path to Purchase Institute.

- For every 4 projects that enter development, only 1 makes it to the market, according to Georgetown University.

It's no wonder why fast-moving consumer goods and retail companies are increasingly listening to what their consumers really want – and using technology to achieve their consumers' dreams.

It is common for consumer testing to occur late in the development cycle. After all, the business case has already been built, and the product design and development completed and incorporated into the company's projections or even at the shelf for in-market testing. Yet many consumers are challenged to see the relevance of new products in a very crowded market. Package design is increasingly understood as an element that can create a powerful consumer experience like no other single element can. Packaging is the thing we see and fall in love with or hate forever. Fast-moving consumer goods and retail companies need a way to create and market packaging and products that will break through the clutter and communicate the brand promise and drive purchases.

Power of the Shopping Experience

What is the shopping experience? Why do you shop the stores you do? Think about it. From the moment you drive into the store parking lot to the moment you leave defines this experience. How easy or difficult is it to park? When you enter the store, what are your senses telling you (look, smell, sound)? What kind of service do you receive? When you search for the items you want to buy, is it easy to find them? The shopping experience, therefore, is the concept of taking the consumers' perspective and understanding their purchasing behavior. The ultimate goal for both the retailer and consumer packaged goods (CPG) company is to design a shopping experience that exactly fits the consumer's wants and needs, allowing them to find what they want when they want it – quickly and easily to increase sales.

This chapter will focus on one element of the shopping experience – the store shelf.

Retailers and CPG companies are eager to know and understand how a consumer decides to buy a product. The difficulty – especially for retailers – is to anticipate this interaction between the consumer and the huge number of products that the brands are delivering. This could rapidly become a nightmare for a consumer who could easily feel lost in front of such a potential visual invasion.

How can retailers and CPG companies guess what consumers will feel and how they will behave in real life? How can they avoid reacting to a downturn in product sales and experiment with changes to the shelf experience in advance? They could invent a time travelling machine to explore the future, or build mock-ups of entire stores and see how shoppers behave. Expensive, isn't it? Or they could experience numerous shelving strategies virtually, linking pragmatism to future exploration.

Delivering a Superior Shopping Experience

When asked what kind of process improvements could be brought to the existing product to create the best consumer experience, professionals such as brand managers, category and space managers, buyers, and C-level executives believe that all retail-driven industries (CPG, CG and retail) could benefit from what many companies have been experiencing with great success over the past 30 years. This includes taking advantage of 3D technology and digital mock ups to visualize future products to make better informed decisions and to collaborate more easily within the enterprise. It can also open the door to enhance the management of a store's operation and scale. Such a virtual universe connects with consumers and enables them to experience the product "in the context" of the store environment. The idea is to establish 3D as the universal language for businesses and consumers to communicate in a way that everyone can understand. This communication would facilitate the transformation of marketing and manufacturing driven industries.

Over the past several years, new technologies have emerged to meet the needs of CPG and retail industries. These technologies make it possible to work on 3D digital objects, all the way from simple packages to a full retail environment. In particular, 3D realistic rendering – that is, the ability to see how lighting, shadows and reflections impact product appearance – is now using advanced real time visualization technologies to provide highly realistic views of stores: shelves, products on them, lighting, building, people, promotional panels, like in the real life. These technologies also come to market with the capabilities inherited of game engines technologies, providing better interactivity and ease of use. The impact of these technologies goes far beyond visualization of 3D virtual products; they also enable global collaboration between multi discipline teams, which can be located virtually anywhere on the planet.

In this context, it is a business solution for brand manufacturers and retailers in the CPG and consumer goods industries that is designed to enable users to simulate retail settings realistically inside immersive, life-like 3D environments to better imagine, validate, and deploy optimum shopping experiences.

Such an application enables category and space managers, buyers, C-level and brand managers to effectively collaborate with their business partners using interactive, 3D virtual stores to review merchandising plans. This solution allows its users to create 3D product representations efficiently from 2D pictures, and then use them to populate virtual shelves and stores, enabling the shopping experience design process to take place inside life-like, 3D environments, significantly faster and at less cost than, for instance, building a real store – only for tests!

Game Changer

As impressive as this technology could be, it would be pointless if they had no potential of transforming the business of consumer products and retail industries. One of the immediate business benefits of these technologies is to move from working on a collection of 2D documents to usage of a unique and online 3D representation of the shopping experience that can be leveraged for variety of connected purposes. 3D is a universal language that enables different people to intuitively interact and understand each other. It enables executives to immediately detect potential problems like space optimization and clash between objects such as store posts, for example. Another potential of the technologies is to support the creation of e-commerce sites that look like the brick-and-mortar store. This would create a better consistency between in-store and online retailing.

Furthermore, these new online technologies are also about better connecting to the end consumers: Seeing everything with their eyes, testing alternatives with them on a variety of technology platforms, from a traditional computer to tablet or inside one of the most advanced virtual reality rooms operated by leading consumer packaged goods and retail companies today.

Finally, the breakthrough of the 3D solution experience would not be as high if it did not include the ability to socially engage in data that was centrally located for all to see, use, re-use and leverage. Information is used once and then "lost" or not easily accessible in CPG and retail companies every single day. People have to redo, re-design, re-explain and re-convince. It can easily drain the life out of an organization's human and financial resources and creates a sub-standard way to compete. A **3DEXPERIENCE** solution is sustained by a platform that enables end users to collaborate effectively from the idea to the execution of the ideal shelf. Information is available whenever they need it, wherever they are around the globe, and always online.

Business Simulator

Things well considered, retail is a very physical, visual and sensitive activity. The importance of this space and its impact on a company's success is gaining more momentum. Yet many are still struggling with disconnected, outdated technology and missing the opportunity to readily enhance the shopping experience. They tend to be sinking in an ocean of figures and reports; old tools do not help them much.

Fortunately, thanks to a new generation of 3D solutions, it is now possible to combine the best of the two worlds: very realistic representations of retail environments and within the same context the capability to smartly navigate on the business reports generated daily by companies. This gives unprecedented visibility on the store environment and thus contributes to take better and faster decisions.

As all types of structured and unstructured information can be displayed, data also connects people that used to work in silos: from sales, marketing, market research, category management, space management, store deployment and senior management. All these different people can look at the retail place with their own view angle while accessing the knowledge coming from their colleagues.

It Has to Be Easy!

Consumers' web sites and mobile devices have shown that user adoption can be huge, but only if the users interface is of compelling simplicity. New tools have to be easy to deploy, easy to use, and easy to emphasize collaboration. They should be useable by anyone in a variety of situations. They should be business accelerators. Drag, drop, create, delete, share – in a few mouse clicks. Also, they should reuse whatever information is accessible to avoid creating duplicate information. For example, modern technologies can leverage existing digital assets like product pictures and dimensions to generate 3D models automatically. They store all data in a central repository that is shared widely throughout the enterprise and externally with partners and suppliers.

Unmatched Precision

Due to the scale, speed and nature of their activity, it is very hard for retailers and consumer goods and CPG companies to deliver precise instructions to the field. This can be dramatically improved with the adoption of 3D technologies. Will the product fit? Can we adapt with a different set of dimensions? What if there is a pillar in this store? All of these questions can be quickly answered by using the most advanced 3D modeling technologies. This makes particular sense when it comes to non-food categories of products, such as leisure, sports and apparel categories. The ultimate precision can be reached. In addition, template technologies can enable all alternatives to be explored quickly and the best compromise can be found for different store formats or nuances. Changes in assortments, available space, and fixtures dimensions are no longer constraints, but become opportunities to differentiate.

Precise and easy to deploy instructions can be generated on the fly and sent to stores. As a result, shelving will be executed "right the first time," compliancy between real stores and specifications will increase, and collaboration between retailers and manufacturers will be improved. The ability to generate clear merchandising instructions is not a process of simplification, but a huge business benefit.

Philippe Loeb is the Consumer Packaged Goods & Retail industry Vice-President for Dassault Systèmes, the 3DEXPERIENCE Company, provider of virtual universes to imagine sustainable innovations. Its solution is designed to transform the way products are designed, produced, and supported. For more information, visit www.3ds.com/CPG.

CHAPTER 16

Making Sense of Social Media

By Michael Schiff

"Can you help me develop my social media plan?" is a question my company is often asked when we begin any engagement with a client. If we're not asked that, we're told, "My social media strategy is doing great. I have 100,000 'Likes' on Facebook." Using that as a measure of social media success is a big pet peeve of mine.

If you're reading this book or work in marketing, you most likely do not need me to tell you how pervasive social media has become. Still, here are a few statistics: A 2010 Nielsen study said the average U.S. consumer each month spends more than sixty hours – or two and half days – online. Of that, nearly 25% is spent on social networks or blogs. Of the 640 million-plus active Facebook users, half log in daily. One hundred and seventy-fve million Twitter users write 95 million tweets – per day. And if you need more proof that social media has truly penetrated our cultural zeitgeist, it has overtaken porn as the #1 activity on the World Wide Web. In fact, according to a 2009 Loyola University Health System study, one in five divorces are now blamed on Facebook.

There are many more statistics out there showing how more and more consumers are engaging with social media. And that is what got me to thinking. Manufacturers and retailers have access to the same statistics I do; maybe some have supplemented that with their own custom research. Still, there are few, if any, case studies they can turn to when putting together their own social media plans. It does not help that the field is changing every day; new vendors are promising to help retailers and manufacturers alike find a way to navigate this ever changing field. We all know how and why consumers are engaging with social media, but very little is known about how the retailers and manufacturers are responding to this new normal.

That is why, over the summer of 2011, my company, Partners In Loyalty Marketing, teamed up with the Shopper Technology Institute to survey leading manufacturers on the role of social media in their marketing strategy. A brief, twenty-question online survey was fielded to 126 manufacturers across a variety of consumer packaged goods (CPG) categories including food/beverage, HBC, and general merchandise. The objective was to establish a baseline of some of the attributes we already know about consumers and their use of social media, including who uses it. We also wanted to learn how companies are using social media for strategy, customer engagement, and how they are measuring it.

So let's take a look at the survey results and figure out what they mean.

Who Uses Social Media?

While most everybody would be hard pressed to find a friend, neighbor or colleague who has not engaged in social media in the past year, they might not think the same thing about CPG companies. Well, the same is true for them as well. Only 11% have stated that social media is not included in their marketing plans; almost 60% participate on both Facebook and Twitter, 30% on Facebook alone.

Interestingly enough, 96% report that their key competitors are using social media. There is somewhat of disconnect between these two numbers. Nearly nine of ten survey respondents (89%) claim that social media is part of their marketing plans, yet believe – rightly or rightly or wrongly – that 96% of the competition is using it as well.

That means there are companies (to the tune of 7% of all surveyed) who, despite the competitive presence, believe that social media is not the place for them. And maybe they are right.

This quote from one of the responders summed up many of the reservations some of the companies have regarding social media: "Some industries are more tightly aligned with social media (tech, media, fashion), but the average grocery store product has little appeal. Do you really want Kraft Mac n' Cheese as your Facebook friend? How eager are you to hear Tweets from Wheat Thins?" In other words, it may make sense and be easy for them, but such a product does not translate very well in the social media arena.

This is an attitude held by many of the companies that do have social media line items in their budget. Over half (56%) of those surveyed said their organization was simply testing the water as far as social media was concerned. They recognize what a juggernaut it is, but are unsure how – and if – they could capitalize on it. Only 13% of all surveyed claimed that social media played a core role in their upcoming business plans and would "for years beyond that." Others saw it as a supporting role, to be used in certain situations – even if they were not quite sure what those situations were.

Category Considerations

This notion of social media benefiting certain product categories over others is an interesting one. To the responder who asked, "How eager are you to hear Tweets from Wheat Thins," we have an answer. There are over 21,000 consumers who are following Wheat Thins on Twitter. Bear Naked, maker of granola products, has over 40,000 fans on their Facebook page. It may be true that certain products lend themselves to the type of engagement promoted by social media. It is equally true that there appears to be a social group for each product out there; they may or may not be mainstream, but they are really into your product. The question then evolves from, "Should I do social media?" to "What do I do with it?"

The Wheat Thins response seems to be to repeat their tagline over and over. Variations of "Crunch is Calling" was littered in every response. Hopefully, if I took the time to follow Wheat Thins on Twitter, I already have a sense that they are indeed crunchy. Granted Twitter only allows 160 characters per Tweet, but I can only believe that these marketers are missing a golden opportunity to further connect with – what must be – a group of their core consumers.

And that might be the biggest benefit of social media to companies. The consumers who get there naturally tend to be among the brand's best or "heaviest" consumers in terms of usage; they are the buyers who make up a disproportionate amount of sales. There is potential to mine these consumers for all sorts of research; for example, a de-facto panel of buyers who can help manufacturers get some insight into what makes these consumers so core to their business. Simply repeating your tagline –something these individuals most likely already know before joining the brand's social media page – is probably not the way to do it.

Bear Naked seems to understand this. A recent view of their Facebook page had shameless promotion alongside consumer insight. In one, they asked consumers to tell them their favorite product and recommended others for them to try. The next post asked their followers to help them find new adventures; they had already done volcano surfing. This post showed an insight into their consumers' collective psyche – information that was most likely procured though other research, but further confirmed on their users Facebook comments.

Budget for Social Media

No matter how brand marketers use social media, one thing is certain: They are using it more and more. Three of four of survey respondents claimed that their company's social media budget grew in 2011 versus 2010; only 1% reported a decline. It still remains, however, a relatively small percent of the overall budget. Two-thirds of the executives surveyed claimed that social media accounted for less than 5% of their total marketing expenditures. This may come as a surprise if one were to rely solely on press releases and media coverage to gauge a company's marketing spend.

Part of the reason the budget may be small for social media is that most companies do not staff against it. Almost 80% of responders claim they have one or less full-time person assigned to social media. In other words – and I say this only partially tongue-in-cheek – they have their interns run their social media campaigns. With one person or less assigned to social media, there can be no comprehensive social media plan. With one person or less, it is difficult to mine one's followers for deeper insights. With one person or less, you have no choice but to be reactive rather than proactive.

Our experience has shown that outsourcing social media to an advertising or public relations (PR) agency can be successful. By allowing them to take care of the program management, the brand can focus on the overall strategy. Plus, an outside agency will typically pay more attention to what is being said, who is saying it, than an overworked, stressed-out brand manager.

Often it is the brand's own perception of their category that drives some of these decisions. "It all depends upon the 'coolness of the brand' or the degree of personal involvement in their decision process. More with beauty care, less with foods" was a direct quote form one survey responder. He was rationalizing his company's low spend on Facebook. He may not be aware that he is also classifying his brand as "uncool." Others take note of category differences, but find ways around them. "Some categories are more blessed with hype and fun, while others are little bit more somber and serious. Regardless, we can still find different ways to engage."

How Do Companies Use Social Media?
Whether an organization is staffing with less than one person or a whole team, once it's decided to use social media as a marketing tactic, the question becomes what to do with it. In the survey, manufacturers were asked that question and their responses fell into one of three buckets:

- *Connect More with Consumers.* "Keep customers and prospects updated about events, news, the market, our portfolio and trends/ developments and to screen 'market moods.'"

- *Build Awareness.* "Build brand loyalty and brand ambassadors."

- *Increase Sales.* "Connecting with consumers and building awareness, which lead to increased sales."

Manufacturers were mostly split between connecting with their existing customer base and building awareness; 41% fell into the former, 48% into the latter category. For the most part, companies are not using social media for e-commerce. In fact only 37% see any potential for social media to drive a direct sale. "We use Facebook to support our products in food, drug and mass and would prefer to have consumers purchase through our customers rather than directly from us. Direct sales . . . is not our core business model," commented one manufacturer. Perhaps the closest manufacturers get to direct sales is with special offers to their fans. Like most offers, these are mostly meant to incentivize customers to go the store to purchase, not for them to necessarily make a purchase online. Companies then have to walk a fine line between attracting deal seekers versus rewarding core buyers.

When we flip things around and asked manufacturers how eager they thought consumers were willing to interact, nearly half (47%) thought that they were very to moderately eager. They bucket consumers into two segments. Some consumers are information seekers. "Consumers want access if they have questions, concerns or just want to share an idea. Social media gives them that access." Others believe that a consumer's eagerness depends on a pre-existing relationship. "Social media is not where an emotional bond is created. The bond needs to extend beyond social media and be pre-existing or it will be an uphill battle."

All of this is used to guide, directly or indirectly, what manufacturers say and do with social media. One thing that is universal – and most manufacturers appear to recognize – is that no matter what you say or tactic you execute, it has to be on the consumer's terms. The danger of mishandling social media is all too real.

Measuring Social Media

This is probably one of the most common questions businesses have today. When our survey asked, "How does your company evaluate the success of social media?" responses fell into one of five buckets:

- Rely on consumer metrics (31%)

- Do not measure at all (22%)

- Look at sales/volume changes (21%)

- Use qualitative and consumer metrics (20%)

- Rely on qualitative measurement only (6%).

Let's take a closer look. One-fifth of businesses have no measurement plan in place whatsoever. They are not even monitoring consumer comments or remarks, which would fall into "qualitative" in the above breakouts.

In fact, only 40% of companies – when pressed – said they analyze social media conversations for insights. Of those, most don't so it regularly. "We do some, but not much" was a common comment. Four of ten responders (43%) said they plan to, but it has been hard to find the time. They have the data and collect the conversations, but are not doing anything with it. Meantime, the pile gets bigger and bigger. The remaining 17% have thrown their hands up in the air and have no intention, "now or ever," of reviewing the different conversations in the social media space. Well, at least they get points for being honest.

Those companies that rely on consumer metrics are really looking at some of the most basic of measures. They consider the number of fans, how often they are going on Facebook/Twitter, and maybe even where they go. "It is extremely hard to measure. There doesn't seem to be a way of isolating the impact from both a volume and brand equity standpoint," said one responder.

And when it comes to ROI, only one of four responders believes it is measurable. "Quantifying the value through ROI and real conversion measurements are barriers and limitations to expanding budgets." I agree, but with a qualifier. I would state that ROI should NOT be the measure used in determining the value of social media. Earlier we stated that most organizations use social media primarily to increase their connection with the consumers and encourage advocacy. One could argue that these are not activities that tie to a direct sale or ROI. Said another way, we may be using the wrong tools to measure social media. It may be better to use more of the same methods that we use to measure awareness or similar objectives.

Measuring Number of Friends

With a lack of any true measurement that ties back to brand health, sales, or overall response to the media in question, companies have resorted to what the average person does when comparing their own social success: They look to see if they have more friends or followers than their own friends do. In the case of companies, they compare themselves to their closest competitors. They then can tell upper management they have 20% more "likes" than the competition. As this is the objective, there is no-shortage of companies and tactics that can help brand accumulate followers to their social platforms. Facebook ads, blogging networks, and digital coupon companies all promise to increase the number of followers a brand has by a significant margin. And they usually succeed.

The problem becomes that – like any popularity contest – as our circle of friends expands, the percent of our tried-and-true core friends begins to diminish. Consequently, their voices begin to get drowned out by the migration of all the followers we actively recruited. A company may start out with only a few hundred friends who are true advocates of the brand. These are the people who searched out the brand on their own and provide the true one-on-one customer feedback that most brands say they want. But then the brand goes out, pays money for a Facebook ad, and recruits 8,000 new fans to their page. That rich dialogue suddenly starts morphing to discussions about deals and when the company will put out an offer.

In fact, when my company delved into the profiles of these new entrants on behalf of our manufacturer clients, we found that they were fans of hundreds upon hundreds of companies. Many of these sites were deal-seeking or coupon sites; others were of companies that were direct competitors of our clients. So much for brand loyalty.

Success of Social Media

In looking at the various responses to the social media survey, I cannot help but conclude that most manufacturers are struggling somewhat with their response to social media. And most would agree. When we reached back out to them and asked how successful social media has been in achieving business objectives, one of three respondents said it was not successful and nearly half (48%) claimed it was only mildly successful. These are not numbers that upper management can usually rally around. Not surprisingly, those who do claim success have the following attributes:

- They have made social media a priority and given it a core strategic role

- It has more than 5% of their marketing budget

- The management of social media is either outsourced or managed by a staff greater than one person

- They are taking steps to measure it – both quantitative and qualitative. It could be true that these organizations started seeing success with social media and then upped their investment – a chicken or egg scenario. But what is evident is that social media is not like most other marketing tactics. If companies don't put money behind it, along with staff and support from upper management, it will be hard pressed to do well.

For a good social media program, it is critical to get organizational fit or buy in from all parties regarding importance and potential. Otherwise brands end up being there for the sake of being there with no clear objective or reason-for-being from the consumers' perspective. In those cases when it is clear that your social media is just not something an organization can get behind or is even worried about, it may be best to invest in other tactics than to invest in a watered-down execution. "As a conservative organization with concerns about controlling the message, this form of media is still too 'risky' from our perspective," comment one manufacturer in the survey.

If an organization does make the plunge into social media, they should do so with some caution. Just because a Facebook page can be put up in an instant does not mean it should. "…not enough effort, investment, and understanding of what we should be doing in social media" was a mantra echoed by many respondents as to why they felt their social media experiences were unsuccessful. Planning should incorporate insuring the right level of resources, usually more than half of a person, and the right communication strategy as well.

Communication is key. This is, after all, social media. If a brand is just there for the sake of being there, then the consumers will only interact when they have a specific need. As one manufacturer commented, "This can be both good and bad . . . another need may be to unload regarding a recent disappointment. This may create a bad impression for everyone." Brands must participate in the discussion and not simply create a forum, which is typically taken over by those who speak the loudest or post most often.

Successful brands claim that they first tried to understand what these consumers needed and wanted and then set up a plan on how to engage with them. "We have found success by interacting with our consumers, posting event photos, working with campus ambassadors and couponing online with retailers," said one manufacturer. Obviously, this is not applicable to every product or business, but creating special experiences and roles for loyal consumers will be critical in a successful execution.

The role of coupons in social media is probably worth exploring a little more. Most companies recognize that to keep consumers excited about their social media experience, there should be the occasional tangible benefit. At the same time, brands don't want to go too extreme, or else risk becoming a haven for deal-seekers. "As long as there is something in it for them, consumers seem to be interested. Once the 'giveaway' or 'offers' wan they move on." It is true that coupons or other similar offers can draw people in, but they can also create an atmosphere of surprise and delight. The challenge is to strike the right cadence of offers. In the end, it is the intangibles that will do most of the heavy lifting. "The level of shopper/consumer engagement is dependent on the perceived value that could come from . . . the image that 'liking' a brand will project."

Intangibles can be any number of things. Those organizations that believe they are having success with social media claim they draw consumers in by being fun and interactive. "Engage with the consumers. They are offering you great feedback that is hard to acquire anywhere else. Be honest, transparent and fun. Make your fans feel valued and important. Don't make it an 'all company/brand' wall or an 'all fan wall. Be interactive." All seemed to follow a similar blueprint. First, be honest. Make everything you do direct and transparent. Next, make it fun. Give consumers games to play, and provide fun and pithy feedback without being condescending. Lastly, provide a balance with content about the company / brand and fans.

While measurement is important, it is equally important to measure the right things. The survey revealed that successful brands claim they were less concerned with ROI and more with monitoring the engagement fans were having. They believed that success in this medium would translate over into sales at the store, and the technology of the tactic may be outpacing the technology in measurement. At least, for now.

Conclusion

It was reported in May 2012 that Facebook is now larger than all of Europe, including Russia. This is not lost on companies. Those that choose to ignore it do so at their own peril, which can be in the form of a consumer creating their own fake site for your brand. Most organizations seem to be dabbling in social media – evidenced by the small amount of budget and resources they dedicate to the medium.

This will need to change. In fact, having a minimal presence on social media will probably end up doing more harm than good. It will become a forum where consumers gather to complain about the product. It is true that different products attract different consumers with different engagement levels. Still, I believe that any brand can offer consumers a mixture of tangible rewards (that is, coupons) and intangible rewards to make them feel valued and appreciative of the brand. If they do that, among consumers who have already shown a fondness of the brand, they can truly foster – if not an emotional connection – certainly increased brand loyalty and perhaps new brand advocates as well.

Waiting for exact measurement is probably not an option. It may be more of an excuse than anything. My company specializes in data-driven strategy and third-party evaluation, so I know how important having the right metrics are for planning purposes. In the end, the objective of this medium is not to directly move units, but to foster the relationship of those people who are most likely the greatest source of brand volume anyway. It is an easy and quick way to connect with core or heavy buyers. What company does not think that is a good idea?

Michael Schiff is managing partner of Partners In Loyalty Marketing (PILM), a Chicago-based consultancy. More information: www.partnersILM.com.

INDEX

Index | 203

3D, viii, 94, 95, 96, 106, 183-7

Account-Specific Marketing, 127, 128
Amazon, 18, 35
American Licorice, 121-2
AMG Strategic Advisors, 125, 131
Analytics, 20, 21, 25, 42, 43, 45, 46, 57, 79, 80, 84, 90, 120, 126, 141
App, 30, 34, 35, 36, 37, 46, 149, 175, 177
Apple, 10
Augmented Reality, 94, 104, 105, 106

Barilla, 145, 154-6
Bear Naked, 191, 192
Best Buy, 34
Big Data, v, ix, 44
Brick-and-Mortar, 30, 31, 34, 37, 84
Brand, 6, 16, 22, 36, 37, 38, 41, 42, 43, 45, 46, 47, 48, 50, 63, 65, 69, 77, 79, 84, 86, 87, 88, 90, 98, 99, 100, 105, 111, 113, 121, 141-9, 151-5, 160, 161, 162, 175-8, 182, 183-4, 192, 193-4, 196-200

Cadbury, 146
Category Captain, 99, 112, 113, 135, 169
Category Management, 65, 88, 90, 98, 99, 135, 186
Cloud, 51, 116, 120
Clustering Technique, 70, 71
Coca-Cola, 34
Compliance, viii, 49, 79, 118, 169-70, 172
Consumer Decision Trees, 83
Consumer Packaged Goods (CPG), 45, 65, 72, 111, 112, 113, 115, 116, 118, 125-9, 131, 133, 135, 171, 182-5, 187, 190
CouponFactory, 179
Coupons, viii, 127, 175-8, 199, 200
Cost of Goods Sold (COGS), 114, 115, 121
CVS, 48, 148

Dassault Systèmes, 187

Decision Insight, 156
Demand Signal Repository (DSR), viii, 133-8
Demographics, 4, 69, 73, 78, 83, 143
Discrete Choice Modeling (DCM), 100
Displays, 35, 78, 84-5, 101, 102, 141, 147
Dunnhumby, 42, 43, 48

E-Mail, 14, 18
Ethnography, 83, 144
Eye Tracking, 46, 94, 102, 144

Facebook, 7, 18, 30, 46, 127, 176, 178, 189-94, 196-8, 200
Fast Moving Consumer Goods (FMCG), 99, 181-2

Gartner Group, 157, 158, 161
GfK, 29, 30, 38
Gladson, 173
Grocery Manufacturers Association (GMA), 47, 98, 157, 183
Google, 35, 105
GPS, 33, 104, 105, 106
Groupon, 18, 175

HALLS, 146
Heineken, 37

IBM, 4, 8, 16, 21, 23
Intel, 95
Internet, 4, 7, 9, 33, 84, 95, 96, 119, 143, 149, 177
Ipsos Marketing, 176
ISI Sharegroup, 169

Kantar Retail, 47, 49
Key Performance Indicators (KPI), 87, 91
Kisok, v, 9
K-Mart, 30
Kraft, 95, 191

Kroger, 42-3, 48, 50, 148, 153
KSS Retail, 165

Land's End, 30
Lego, 31
LinkedIn, 176
LivingSocial, 18, 175
Location-Based Services, 33, 106
Loyalty Card, v, 49, 80, 86

McKinsey, 44
Merchandising, 21, 41, 42, 46, 55, 65, 72, 78, 79, 83, 84, 85, 91, 94, 99, 101, 125, 126, 147, 151, 153-4, 157, 158, 163, 169-72, 184, 187
Mobile Device, 3, 18, 30, 35, 186
Moment of Truth, 58, 77, 81, 86, 90
Multigenerational Household, 12

National Association of Retail Marketing Services (NARMS), 172
Nestlé, 145, 146, 151-4
Neuroscience, 104, 144
Nielsen, 134, 189
Nike, 31

Out of Stocks, viii, 18, 42, 136, 157, 158, 160, 164, 170, 172

Packaging, 84, 90, 100-1, 104-5, 142, 143, 147, 149, 150, 151, 152, 182
Partners in Loyalty Marketing, 190, 200
Path to Purchase, v, ix, 27, 43, 45, 77, 78, 80, 81, 84, 88, 89, 90, 94, 141, 150, 154
PepsiCo, 95
Planogram, viii, 36, 37, 83, 93, 98, 100, 146, 148, 154, 156, 169-72
POPAI, 45
POS, viii, 85, 89, 101, 102, 103, 105, 133-5, 138, 158, 160, 161
Precima, 56, 63
Pricing, 17, 46, 54, 81, 83, 84, 88, 100, 103, 120, 125, 128, 143, 147, 148, 150, 151, 153

Procter & Gamble, 7, 46, 95

Ready-for-Purchase (RFP), viii, 158-61, 163-5
Relational Solutions, 139
Robinson-Patman Act, 50
ROI, 84, 88, 89, 90, 116, 127, 135, 147, 150, 196, 199

Saas, 116, 118, 119, 120, 138
Safeway, 48, 153
Sears, 30
Share of Shopper, 85, 86, 87
Shelf Set, 84, 146, 147
Shopper Behavior, 32, 43, 45, 46, 72, 73-4, 78-9, 83-4, 89, 90, 101, 141, 143, 151
Shopper-Centricity, vii, 53-8, 63, 67, 127
Shopper Data, 41, 42-4, 49, 54, 77, 79, 88
Shopper Insights, 41, 43, 45, 47-9, 54, 66, 77, 98, 99, 152, 155
Shopper Marketing, 27, 31, 32, 38, 43-5, 47-9, 88-90, 98, 99, 104, 125-8, 130, 142, 145, 147, 150, 151, 154
Shopper Metrics, vii, 55, 77, 79, 81, 84, 88, 89, 90
Shopper Segmentation, vii, 42, 65-70, 72-3
Shopper Technology Institute, ix, 190
Showrooming, 31
Skinny Cow, 151, 153-4
SKU, 113, 121, 129, 146, 149, 152, 153, 155, 171
Slotting, 125, 129
Smartphone, v, 7, 27, 28, 29, 30, 33, 35, 104, 177
Social Media, v, ix, 6, 9, 17, 22, 30, 35, 41, 46, 127, 136, 141, 176, 189-200
Spire, 73, 74
Starbucks, 36
Sunny D, 120-1

Tablet, v, 28, 29, 30, 33, 184
Trade Promotion Management (TPM), vii, 111, 114, 117-23, 128, 138

Trade Promotion Optimization (TPO), 116-7, 120, 128, 138
Transaction Data, 41, 44, 83
Target, 45, 50, 51, 148
Tesco, 42
TradeInsight, 120, 122, 123
Transaction Data, 41, 44, 79, 83,
Twitter, 5, 7, 44, 127, 176, 189, 190, 191-2, 196

Validation, 98, 143, 150
Video Analytics, 78, 84-6, 90
VideoMining, 78, 83, 91
Virtual Reality, 46, 94, 96, 104, 106, 185
Virtual Shopping, vii, viii, 93, 95-102, 104, 142-51, 153-4
Vision Critical, 97, 98, 99, 105, 107
Vodafone, 36, 37

Walgreen's, 148
Walmart, 35, 36, 46, 47, 50, 51, 135, 141, 148
Wellness Natural Food, 34
Wheat Thins, 191-2
Williams-Sonoma, 34

Xtreme Shoppers, 28-9, 38

YouTube, 7, 9